Silicon Carbide One-dimensional Nanostructures

FOCUS SERIES

Series Editor Michel Pons

Silicon Carbide One-dimensional Nanostructures

Laurence Latu-Romain
Maelig Ollivier

iSTE

WILEY

First published 2015 in Great Britain and the United States by ISTE Ltd and John Wiley & Sons, Inc.

ISTE Ltd
27-37 St George's Road
London SW19 4EU
UK

www.iste.co.uk

John Wiley & Sons, Inc.
111 River Street
Hoboken, NJ 07030
USA

www.wiley.com

Library of Congress Control Number: 2014955862

British Library Cataloguing-in-Publication Data
A CIP record for this book is available from the British Library
ISSN 2051-2481 (Print)
ISSN 2051-249X (Online)
ISBN 978-1-84821-797-3

Contents

Foreword

Silicon carbide (SiC) is a wide bandgap semiconductor having factors of merit scaling well above those of well-established semiconductors such as silicon (Si), which became the material of choice of the 20th Century. SiC is a ceramic with very interesting mechanical properties (hardness and composite materials) and a very high thermal conductivity, just like copper. In addition, due to its chemical inertness and its ability to resist radiation damage, this compound semiconductor is an especially suitable electronic material to be used in harsh environments. Furthermore, SiC is also biocompatible, making it promising for interfacing an electronic material with biology. SiC is also very abundant in the interstellar medium and could be a key material in the route toward prebiotic life apparition in the universe. Indeed, SiC was found by Henri Moisan at the end of the 19th Century on a meteorite fallen in the Diablo Canyon, Arizona. The overall properties of SiC have triggered many advanced applications such as high-temperatures, high-power, high-frequencies devices and sensors. Finally, SiC is also an especially interesting material in nanoscience and nanotechnology, such as in the self-organization of massively parallel Si passive/active atomic lines at its surface, or of carbon nanotubes, and an excellent substrate for the epitaxial growth of high-quality graphene. At this point, it is appropriate to mention that the research programs on Si nanowires resulted in about 9,000 publications since the year 2000, while in contrast, similar research on SiC remains rather marginal with no more than 400 publications, despite the much higher performances of SiC when compared to corresponding Si nanowires – see the Introduction section.

In this context, the present book, written by Latu-Romain and Ollivier, brings novel insights in describing a very original and innovative approach to grow SiC nanotubes and nanowires. Indeed, at variance to the well-established methods to grow nanowires, the idea presented here is to grow SiC nanowires/nanotubes using Si nanowires grown through a top-to-bottom approach as an initial template. Then, selective carburization of the external surface core–shell leads to SiC growth covering the outer surface of these Si nanowires. Subsequently, the carburization process leads also to Si removal by out-diffusion, leaving only high-quality SiC nanotubes. These one-dimensional nano-objects potentially open exciting new applications not only in electronics or sensors, but also toward biology and medical applications such as selective drug delivery.

In this book, the authors start with an introduction, and then all the documents are organized into four different chapters, ending up with a conclusion. The Introduction presents the topic of interest, with a short historical background, putting issues and challenges into perspective. Chapter 1, entitled "*Properties of SiC Based One-Dimensional Nanostructures*", provides a thorough presentation of the SiC crystallography, describing the structure of some of the most important SiC polytypes (out of more than 170). It includes cubic 3C, hexagonal 2H, 4H and 6H, and rhomboidal 15R. They also present a general view of their physical, chemical and electronic properties, which are key aspects for understanding SiC, especially at the nanometric and subnanometric scales. Moving next to Chapter 2, entitled "*State of the Art of the Growth of SiC-Based One-Dimensional Nanostructures*", the authors address the key points that make the central issues of their book, namely the growth techniques and challenges. The first section describes the state of the art growth of SiC nanowires. The second section is devoted to the state-of-the-art growth of SiC nanotubes, while the third and last sections move to state-of-the-art growth of SiC-based core–shell nanowires. Chapter 3 is entitled "*An Original Growth Process: the Carburization of Si Nanowires*" and presents the novel approach and key methods that have been developed to do so. It is organized in six sections, respectively, describing Si nanowires, the carburization of bulk silicon, experimental applications, the growth of core–shell Si–SiC nanowires and of SiC nanotubes, and a summary of the study of the carburization of silicon nanowires. Finally, in the last chapter, Chapter 4 entitled "*SiC-Based One-Dimensional Nanostructure Technologies*", the authors describe some promising potential or already existing novel applications. The top-down approach using SiC plasma etching for SiC nanowires is described in the first section, while the mechanics and potentially subsequent novel composites are addressed in the second section. In the third section, energy applications, such as energy storage through super-capacitors, are envisioned, while in the fourth section, novel electronics applications, such as SiC nanowire transistor, are proposed and described. Finally, biology applications and future works are described and commented upon the fifth and six sections, respectively. In the Conclusion, the authors summarize the know-how and the subsequent results for these one-dimensional SiC nanostructures and draw perspectives toward future promising applications covering various high-tech domains of science and technology.

We can therefore envision that this book is providing a very suitable document based on sound grounds that will be especially useful for scientists and engineers performing research and developing technological applications. Furthermore, this same book will also serve as a very appropriate, pertinent and well-focused tutorial for graduate master as well as for PhD students.

Patrick SOUKIASSIAN
Commissariat à l'Energie Atomique et aux Energies Alternatives, Saclay, France
and
Synchrotron SOLEIL
Saint Aubin, France
November, 2014

Introduction

Increasing communication requirements, as well as the need for processing of data or diagnostics in general, which we have experienced for several decades now, have allowed several sectors, such as microelectronics, information technology (IT) or even telecommunications and biotechnology to become strategic, globalized and recognized industrial sectors. These different technological domains have greatly benefited from scientific progress, notably in the domains of materials and microelectronics.

The discovery of semiconducting materials, for example, materials whose behavior with regard to current can be modified by adding impurities, is definitely one of the advances that allowed for the revolutionizing of most of the domains named above. Among these materials, silicon has become an emblem, to the point of becoming nearly hegemonic. Despite this, its physical properties limit its use to certain sectors such as optoelectronics, or hinder the obtaining of satisfactory results, like the low yields of solar cells, for example.

On top of the materials-related physical limitations, there are also technological limitations, when for reasons of space, yield or mobility, new technologies reach nanometric sizes.

This scale change has brought new challenges for those involved in the industry and researchers, the main ones being the creation of nanometric objects and the control of the physical and chemical properties of these objects. This book fits into the general scope of this challenge.

One-dimensional (1D) nanostructures, such as nanowires or nanotubes, are of interest to the scientific community with regard to their growth, characteristics and their potentially exceptional properties compared to the bulk material. Semiconducting 1D nanostructures are widely studied, mainly those made of silicon and III-V materials for their electronic and optic properties.

In this book, we will look at the unique case of the growth of 1D nanostructures of silicon carbide (SiC). In the literature, this subject has not been explored greatly as the synthesis of nanostructures is difficult. The polytypism of SiC – and therefore its structural stability – is indeed an additional difficulty in the growth of such structures.

Figure I.1 illustrates the growing interest of researchers for semiconducting 1D nanostructures since the 2000s. The number of publications recorded in *Isi Web of Knowledge,* which include the keywords *Si nanowire* and *SiC nanowire,* is shown. The number of publication reaches 9,000 in the first case and only 400 in the second case. Only half of these works are on the study of the growth of these nanostructures. Finally, the studies related to the synthesis of SiC nanowires are 20 times less common than those on silicon nanowires. However, we will see that the unique properties of SiC merit that time be spent on the study of the growth of these nano-objects. The different techniques used in the synthesis of SiC nanowires and nanotubes are reviewed. We have also elected to study an original growth process based on the carburization of SiC nanowires. This approach has allowed for work on the growth of SiC 1D nanostructures from the excellent base that is the single crystalline silicon nanowire. Original nanostructures, such as Si-SiC core-shell nanowires and SiC nanotubes, have thus been obtained. The different application domains for mechanics, energy, electronics and the biology of the use of these structures are finally described.

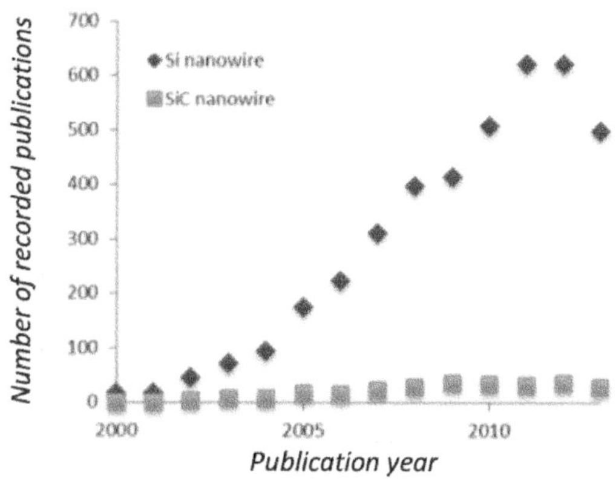

Figure I.1. *Number of publications recorded in Isi Web of Knowledge by the year from 2000 to 2013 containing the words Si nanowire and SiC nanowire (statistics carried out December 9th 2013)*

List of Acronyms

1D	One-dimension
CVD	Chemical vapor deposition
FET	Field Effect Transistor
FIB-SEM	Focused ion beam-scanning electron microscope
HR-TEM	High-resolution transmission electron microscope
IT	Information technology
JFET	Junction Field Effect Transistor
LA	Longitudinal acoustic
LO	Longitudinal optic
MEMS	Micro-electro-mechanical systems
MOSFET	Metal Oxide Semiconductor Field Effect Transistor
NS	Nanostructure
NT	Nanotube
NW	Nanowire

SEM	Scanning electron microscope
SF	Stacking fault
SGPS	Solid and Gases Pure Substances
SiC	Silicon carbide
STEM	Scanning transmission electron microscope
TA	Transverse acoustic
TEM	Transmission electron microscope
TMS	Tetramethylsilane
TO	Transverse optic
VLS	Vapor-Liquid-Solid
VS	Vapor-Solid
VSS	Vapor-Solid-Solid

1

Properties of SiC-based One-dimensional Nanostructures

This chapter is based around the presentation of the physical properties expected in silicon carbide (SiC)-based one-dimensional (1D) nanostructures. First, the properties of the SiC material will be presented, focusing on the electronic properties of SiC as well as its crystalline properties. Next, a definition of 1D nanostructures and a demonstration of their physical and chemical properties will be put forward.

1.1. Intrinsic properties of silicon carbide

1.1.1. *Crystallographic description*

1.1.1.1. *The tetrahedron, base unit of SiC*

Based on the Si—C bond, with a length of $d_{Si—C}$ = 0.189 nm, a bond that is both covalent (89%) and ionic (11%), SiC is organized in the form of tetrahedrons (see Figure 1.1[1]) with a carbon atom (or silicon atom) at their center, linked to four silicon (or carbon) atoms. In such a

1 The crystallographic representations were carried out using the Visualization for Electronic and Structural Analysis (VESTA) program [MOM 11].

configuration, the nature of the atoms at the summit of the tetrahedron and in the tetrahedral site is interchangeable. However, for more clarity, all of the explanations will be given with one single configuration: the atom of C in a tetrahedral position. This tetrahedral organization is easily explained if we consider the electronic structures of the silicon and carbon and the hybridization of the atoms in the lattice. These two atoms are part of column IV of the periodic table of elements and have the electronic configurations of [Ne] $3s^2$ $3p^2$ and [He] $2s^2$ $2p^2$ for silicon and carbon, respectively. In the SiC lattice, the atoms are hybridized into sp^3 allowing the valence electrons to form four bonds. The angle between two Si—C bonds is then equal to $\alpha = 70.529°$. Let us note that in this configuration, the basal plane consisting of three silicon atoms and the next plane of silicon atoms are separated by $d_{\text{plane Si–plane Si}} = 0.252$ nm, while the plane containing the carbon atom in the tetrahedral interstitial space is $d_{\text{plane Si–plane C}} = 0.063$ nm from the basal plane.

The tetrahedrons can be orientated in two ways by a 60° rotation around the $\vec{c}$ axis of the hexagonal system: we talk of "right" or "left" tetrahedrons. The $\vec{c}$ axis of the hexagonal system corresponds to the axis [111] of the cubic system and axis [0001] of the hexagonal system (see Figure 1.2).

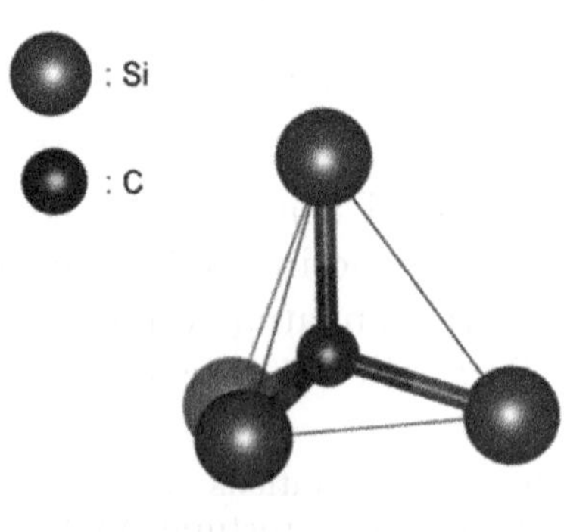

Figure 1.1. *Representation of the characteristic SiC tetrahedron. The smaller carbon atom has an interstitial position*

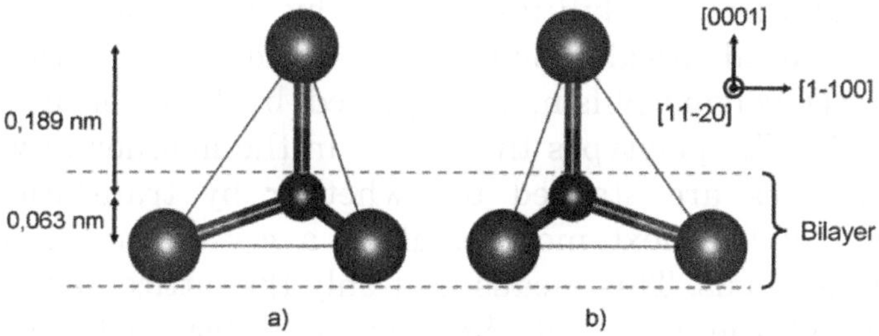

Figure 1.2. *Schematic representation in the (11$\bar{2}$0) plane of the a) "left" and b) "right" tetrahedrons, along the vertical axis [0001]. A bilayer, made up of the two first planes of atoms is also represented, as well as the distances $d_{plane\ Si-plane\ C}$ and d_{Si-C}*

Thus, SiC can be represented as a stack of this tetrahedral unit by their summits. From this stack comes one of the particularities of SiC, called polytypism. Indeed, SiC can be organized into different crystalline networks – the polytypes – according to the manner in which the tetrahedrons are stacked up along the $\vec{c}$ axis.

1.1.1.2. *Polytypism of SiC and Ramsdell notation*

The polytypism phenomenon has been particularly studied in the case of SiC, due to the large number of polytypes – more than 200 – which have been recorded up to now. By the middle of the 20th Century, the scientific community had identified four distinct SiC polytypes, and the crystallographic study of these has been the subject of several scientific publications, notably by Baumhauer [BAU 15], who reported for the first time the existence of several SiC polytypes, or even Thibault who in 1944 referenced all of the works carried out on this particularity of SiC [THI 44]; but also Ramsdell [RAM 47] and Shaffer [SHA 69] who contributed to a better understanding of SiC polytypism.

Polytypism is the ability of a substance to crystallize into different crystalline structures. Polytypism is a particular case of 1D polymorphism, since it is limited to the crystallization possibilities offered by the stacking of a "module" of identical chemical composition. This is referred to as 1D polymorphism, as reported by Schneer in 1955 [SCH 55]. The polytypes thus differ in the manner in which the modules are stacked up, whether by translation or rotation of the next module, along a crystalline direction along the $\vec{c}$ axis. This means that only the lattice parameter outside of the plane of the module differs between the polytypes.

In the case of SiC, the base "module" is made of a bilayer formed by the basal plane, itself made of Si atoms, and the plane containing the carbon atom in the tetrahedral site (see Figure 1.2). This bilayer can be replaced by the tetrahedron, since the tetrahedrons are joined in their summits. Let us consider the ideal case where the base of a SiC crystal is made of a single type of tetrahedron, the "left" tetrahedrons (see Figure 1.2). The next layer of tetrahedrons, depending on the $\vec{c}$ axis, can be made either of "right" tetrahedrons or "left" tetrahedrons. The third layer can also be found to be made of these two types of tetrahedron, and so on and so forth until a complete crystal is formed. The result is an infinite amount of possibilities for the SiC to crystallize.

The different polytypes are identifiable using the Bravais lattices. SiC crystallizes mainly as hexagonal, rhombohedral or cubic systems. However, there can be several polytypes associated with one of the Bravais lattices.

Historically, the different polytypes were, therefore, named by number, corresponding to their chronological order of discovery. To be able to offer a more scientific nomenclature, Ramsdell proposed in 1947, in an article called *Studies on silicon carbide* [RAM 47], to use the crystallographic data associated with each polytype. The notation proposed by Ramsdell, and which carries his name today, is the following: the first letter of the Bravais lattice in which the structure is crystallized (C for cubic, H for hexagonal and R for rhomboidal) is coupled with the number of bilayers needed to entirely describe the structure. The most common polytypes are thus noted:

– 3C where all the tetrahedrons are orientated in the same way (see Figure 1.3);

– 2H where the "left" and "right" tetrahedrons alternate (see Figure 1.4);

– 4H polytype corresponds to the stacking of two "left" tetrahedrons" followed by two "right" tetrahedrons (see Figure 1.5);

– 6H polytype corresponds to the stacking of three "left" tetrahedrons followed by three "right" tetrahedrons (see Figure 1.6);

– 15R polytype corresponds to the stacking of three "left" tetrahedrons followed by two "right" tetrahedrons, this pattern being repeated three times (see Figure 1.7).

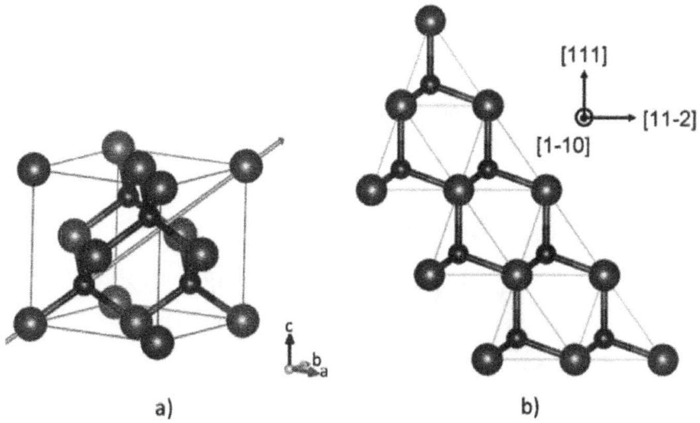

Figure 1.3. *Schematic representation of the 3C–SiC polytype, which corresponds to the zinc blende structure. a) Primitive cell of 3C–SiC. We can observe the carbon atoms in the interstitial position. [111] axis of the cubic system, which corresponds to [0001] axis in the hexagonal system, is represented by an arrow. b) Stack along axis [111] of the tetrahedrons describing the structure of the 3C–SiC polytype in plane ($1\bar{1}0$)*

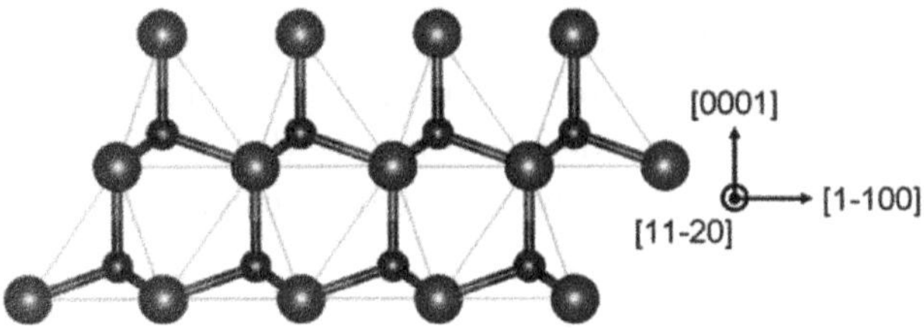

Figure 1.4. *Schematic representation of stacking along [0001] axis of the tetrahedrons describing the structure of 2H polytype in $(11\bar{2}0)$ plane*

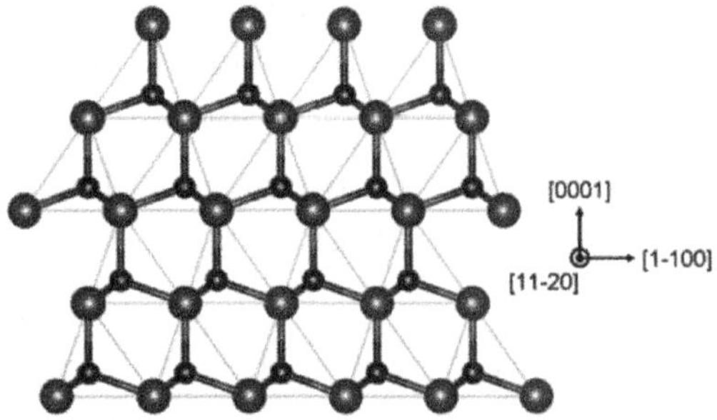

Figure 1.5. *Schematic representation of stacking along [0001] axis of the tetrahedrons describing the structure of polytype 4H–SiC in $(11\bar{2}0)$ plane*

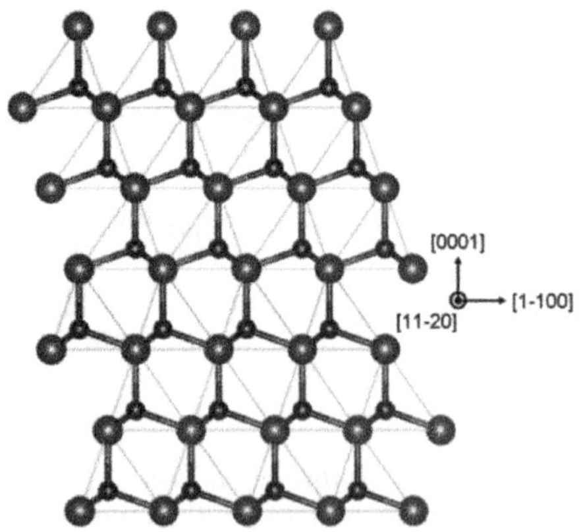

Figure 1.6. *Schematic representation of stacking along [0001] axis of the tetrahedrons describing the structure of polytype 6H–SiC in $(11\bar{2}0)$ plane*

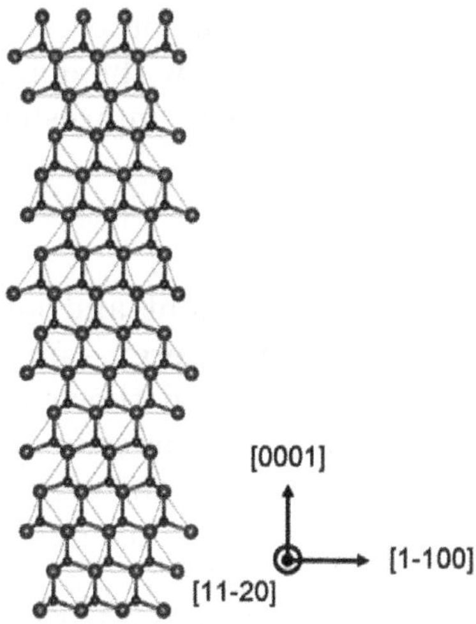

Figure 1.7. *Schematic representation of stacking along the [0001] axis of the tetrahedrons describing the structure of polytype 15R–SiC in the $(11\bar{2}0)$ plane*

Table 1.1 lists the crystallographic data of the main polytypes of SiC. It must be noted that only one cubic form exists: 3C–SiC. To distinguish it from the other polytypes, it is sometimes written β–SiC, as opposed to other polytypes listed under α–SiC.

Polytypes		*3C*	*2H*	*4H*	*6H*	*15R*
Space group		$F\bar{4}3m$	$P6_3mc$	$P6_3mc$	$P6_3mc$	$P6_3mc$
Lattice parameters	*a* (Å)	4.359	3.081	3.081	3.073	3.073
	c (Å)	–	5.031	10.061	15.08	37.70

Table 1.1. *Crystallographic data of the main polytypes of SiC, from [MOR 94, BEC 97, LAT06]*

1.1.2. *Physical and chemical properties of SiC*

The exceptional properties of SiC are the main reason behind the research and development of this material. In this section, first the main physical properties of SiC are presented, and then the electronic properties of this material are developed, so as to explain how the domain of micro- and nano-electronics is particularly interested in SiC as a semiconductor.

1.1.2.1. *General properties*

As a result of the both covalent and ionic nature of the Si—C bond, as well as its high energy – 5 eV – and the small distance d_{Si-C}, SiC possesses good thermal and mechanical properties, and resists well to corrosive agents. Thus, the thermal conductivity of SiC is slightly greater than that of copper (copper = 3.9 $W.cm^{-1}.K^{-1}$), the high Young's modulus and the hardness of SiC make it an ideal material for abrasion; finally, it is resistant to ionizing radiation, acid or base etching, and does not oxidize easily[2]. Table 1.2 compares the properties of SiC with other semiconductors: silicon (Si), the diamond form of carbon, aluminum nitride (AlN) and gallium nitride (GaN). This comparison shows that SiC possesses properties close to those of AlN and GaN, better than those of silicon, but not as good as diamond. However, three points must be retained:

– first of all, while diamond is indeed better for mechanical applications (abrasion, material cutting, etc.), it is harder and therefore more expensive to make, as reflected by the fact that synthetic SiC is often used as a diamond imitation;

– the biocompatibility of SiC, coupled with its hemo-compatibility, makes it different from all the other semiconductors and opens up a field of future application [SAD 11a, SAD 11b];

– finally, while many properties are common to all the polytypes, we must remind ourselves that 3C–SiC is the only

2 SiC has the same native oxide as Si, which is SiO_2. This shared point is a definite advantage for microelectronics.

isotropic crystallographic form, and this obviously has consequences with regard to its properties, in the case of a single crystal.

In the end, it appears that SiC possesses good properties of thermomechanical applications when considered as a ceramic. Moreover, its good thermal properties and biocompatibility, coupled with its electronic properties, make it an ideal candidate for micro- and nano-electronics as well as for biotechnology.

<table>
<tr><th>Materials</th><th>Melting point (F) or non-congruent melting point (S) (°C)</th><th>Thermal conductivity λ at room temperature ($W.cm^{-1}.K^{-1}$)</th><th>Isotropy</th><th>Young's modulus (GPa)</th><th>Density ($g.cm^{-3}$)</th><th>Resistance to chemical etching</th><th>Biocompatibility</th></tr>
<tr><td>3C–SiC</td><td rowspan="5">2,200 (S)</td><td rowspan="5">4.5</td><td>Yes</td><td rowspan="5">450</td><td rowspan="5">3.21</td><td rowspan="5">Very good. Acid and basic</td><td rowspan="5">Yes. Bio- and hemo-compatible</td></tr>
<tr><td>2H–SiC</td><td>No</td></tr>
<tr><td>4H–SiC</td><td>No</td></tr>
<tr><td>6H–SiC</td><td>No</td></tr>
<tr><td>15R–SiC</td><td>No</td></tr>
<tr><td>Si</td><td>1,414 (F)</td><td>1.5</td><td>Yes</td><td>150</td><td>2.33</td><td>Moderate for acids, weak for bases</td><td>No</td></tr>
<tr><td>C diamond</td><td>> 3,800 (S)</td><td>20</td><td>Yes</td><td>1,000</td><td>3.51</td><td>Excellent</td><td>Yes</td></tr>
<tr><td>AlN</td><td>> 2,500 (S)</td><td>2.5</td><td>Depends on the phase</td><td>310</td><td>3.20</td><td>Very good</td><td>No</td></tr>
<tr><td>GaN</td><td>2,200 (F)</td><td>0.8–1.3</td><td>Depends on the phase</td><td>150–180</td><td>6.15</td><td>Very good</td><td>Yes</td></tr>
</table>

Table 1.2. *Comparison of some physical and chemical properties of SiC with different materials: Si, diamond, GaN and AlN. From [SAD 11a, RAY 07, CAM 98, MER 09, NAL 02]*

1.1.2.2. *Electronic properties*

As SiC is composed of atoms belonging to column IV of Mendeleev's periodic table, it is itself a semiconductor. It is associated with large bandgap semiconductors, and the value of the energy of the bandgap varies as a function of the polytype between 2.2 and 3.3 eV. Table 1.3 groups the electronic properties of the main polytypes of SiC[3] as well as those of the semiconductors Si, Ge, C (diamond), GaN, AlN, GaAs and InP. The energy of the bandgap E_g, the saturation velocity of the electron sv_{sat}, the breakdown field E_B, the mobility of the electrons' and holes' mobility, μ_n and μ_p, the relative permittivity ε_r, the thermal conductivity λ and the maximal operating temperature T_{max} are reported. From this table, it is possible to extract several points that make SiC interesting:

– the indirect bandgap value is high compared to that of silicon, GaAs and Ge. This value, therefore, limits the use of SiC for low-consumption applications, where it is not adapted. However, with such a bandgap, leakage currents at high temperatures are strongly limited. Indeed, the intrinsic carriers will need more energy to pass spontaneously in the conduction band;

– the fact that each polytype possesses a different value of E_g also allows the consideration of a greater field of application, since it is possible to reach a whole range of E_g;

– the indirect nature of the bandgap of SiC – just like for silicon – limits its use in optoelectronics, despite the fact that such applications have already been explored [BRO 93, HU 08, BAI 07]. However, SiC acts as a seed for the epitaxial growth of GaN, a material that is widely developed for

3 It must be noted that the electronic properties of 15R–SiC have not yet all been measured, despite the fact that it can be grown.

optoelectronics. Indeed, the lattice parameter a of GaN, crystallizing in a hexagonal system ($a = 3.186$ Å and $c = 5.186$ Å) is very close to the parameters a of the polytypes 4H and 6H of SiC (see Table 1.1);

– the saturation velocity of electrons v_{sat} corresponds to the maximal speed at which the electrons can move when they are submitted to a strong electric field. This speed is higher in SiC compared to most other semiconductors. Moreover, as reported by Morkoç *et al.* in 1994 [MOR 94], SiC does not experience any decrease in v_{sat} for strong electric fields, but only a threshold (see Figure 1.8). Moreover, we can note that polytype 3C–SiC possesses a higher v_{sat} than the other polytypes of SiC, due to a reduced diffusion of the phonons;

– the high values of the breakdown field (E_b), as well as the relatively good mobilities, while lower than those of Si, of the electrons (μ_n) and of the holes (μ_p) of the SiC polytypes, explain the use of SiC for high power applications. Indeed, coupled with the correct thermal conductivity of SiC and with its maximal operating temperature, these properties make SiC ideal to allow the flow of a strong current, without overheating the component. These properties have long been exploited to make Schottky diodes functioning at up to 600 V and 60 A (Cree® company [AGA 13]), Junction Field Effect Transistor (JFET) and Metal Oxide Semiconductor Field Effect Transistor (MOSFET) transistors (the Cool SiC™ 1,200 V SiC JFET Family from Infineon® components), as well as thyristors able to withstand up to 6,500 V, for instance.

It would, however, appear that carbon in the diamond form usually possesses better electronic properties than SiC. While this is true, this affirmation meets several difficulties.

There is no complete technological stream based on diamond, due to the current impossibility to obtain good quality and large substrates (substrates greater than 100 mm). Moreover, diamond is easily oxidized when heated to more than 400°C by forming CO_2 and CO, which limits its use for high temperature applications. These two points are resolved in the case of SiC, since commercial substrates are already on the market (4H and 6H), and SiC is difficult to oxidize at high temperatures, due to the formation of a protective layer of SiO_2. With regard to GaN, it is important to recall the scarcity of global gallium resources compared to silicon and carbon resources, as well as the fact that currently there are no commercial free-standing GaN substrates available.

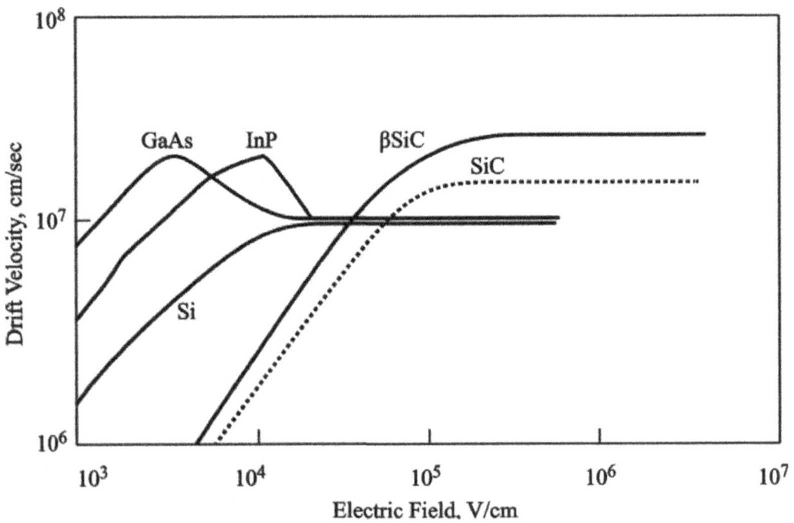

Figure 1.8. *Graph representing the saturated electron velocity v_{sat} as a function of the applied electric field. From Morkoç et al. [MOR 94]*

Materials	*Indirect bandgap value E_g(eV) direct (d) or indirect (i)*	*Speed of the electrons at saturation v_{sat} ($\times 10^7$ cm.s^{-1})*	*Breakdown field E_B ($\times 10^5$ V.cm^{-1})*	*Electron mobility μ_n and hole mobility μ_p (cm^2.V^{-1}.s^{-1})*	*Relative permittivity ε_r*	*Maximal operating temperature T_{max} (°C)*	*Thermal conductivity λ at room temperature (W.cm^{-1}.K^{-1})*
3C–SiC	2.2 (i)	2.5	15	μ_n = 1,000 μ_p = 40	9.6	925	4.5
2H–SiC	3.3 (i)	2.3	?	μ_n =1,500 μ_p= ?	9.7	? ≈ 1000	
4H–SiC	3.2 (i)	2	20	μ_n = 1,000 μ_p = 115	10	1450	
6H–SiC	2.9 (i)	2	22	μ_n = 800 μ_p = 90	9.7	1300	
15R–SiC	3.0 (i)	?	?	?	?	? ≈ 1000	
Si	1.12 (i)	1.1	2.5	μ_n = 1,400 μ_p = 370	11.8	325	1.5
Ge	0.66 (i)	0.5	1	μ_n = 3,900 μ_p = 1,900	16	300	0.6
C (diamond)	5.5 (i)	2.7	100	μ_n = 2,000 μ_p = 2,100	5.6	560	20
GaN	3.3 (d)	2.5	14	μ_n = 900 μ_p = 150 ?	11.1	1650	1.3
AlN	6.2 (d)	2.1	12–18	μ_n = 300 μ_p = 14	8.5	? $T_{melting}$ = 2200	2.85
GaAs	1.4 (d)	1.0	3	μ_n = 8,500 μ_p = 400	12.8	490	0.5
InP	1.34 (d)	1.0	5	μ_n = 4,600 μ_p = 150	12.4	? $T_{melting}$ = 1062	0.68

Table 1.3. *Comparison of some electronic properties of the main SiC polytypes with different semiconductors: Si, Ge, diamond, GaN, AlN, GaAs and InP. From [LAT 06, RAY 07, CAM 98, MER 09, NAL 02, FER 06, MOH 96, NIL 91]*

From the exposition of the physical and chemical properties of SiC, it is possible to extract some important information:

– SiC possesses several crystalline forms, called polytypes, coming directly from crystallographic considerations via the stacking of the SiC_4 tetrahedrons (or CSi_4). Three polytypes do stand out, however, by their possible growth: the cubic 3C–SiC form, and the hexagonal 4H–SiC and 6H–SiC forms;

– the physical properties of SiC come from the Si–C bond, while maintaining the idea of polytypism. The excellent thermal properties of SiC, as well as its high resistance to harsh environments, (irradiation, acidic or basic media and high temperatures) make it an exceptional material;

– the biocompatibility, as well as the hemo-compatibility, of SiC places it in a category of its own among semiconductors and provides new possibilities;

– from an electronic point of view, in the application fields such as power electronics and/or functioning in a harsh medium, SiC appears as the ideal material due to its high resistance, good thermal properties and ability to withstand strong currents and voltage, where silicon is not suitable;

– globally, 3C–SiC is the polytype with the best electronic properties, notably due to its isotropy, greater electronic mobility and the smaller value of its bandgap.

1.2. Properties of one-dimensional nanostructures

1.2.1. *Definition and classification*

Before presenting the different physical and chemical properties of 1D nanostructures, it is necessary to define the latters. When looking at nano materials, the legal definition proposed by the European Commission and quoted hereafter provides an initial legal framework, notably in terms of dimensions: a nano material is "a natural, incidental or manufactured material containing particles, in an unbound state or as an aggregate or as an agglomerate and where, for 50% or more of the particles in the number size distribution, one or more external dimensions is in the size range 1 nm–100 nm"[4].

1D nano-objects can, therefore, be defined as:

> *One-dimensional nano-objects are any three-dimensional objects for which at least two dimensions are smaller or equal to 100 nm.*

Using this concise definition, it is possible to group a large number of terms that appear gradually as new scientific discoveries are made. However, we can distinguish two large categories of 1D nano-objects: "hollow" 1D nano-objects and "full" 1D nano-objects.

1.2.1.1. *"Hollow" 1D nano-objects*

Usually called nanotubes, these objects are made of walls surrounding an empty space. The wall can be composed of a single atomic plane: these are single-walled nanotubes; or they can be made of several atomic layers (multi-walled nanotubes). Among these nanotubes, the best known, most studied and most used are without a doubt carbon nanotubes. Discovered by Iijima, who published his work in *Nature* in 1991, carbon nanotubes are made of layers of graphite bent over themselves [IIJ 91, END 97]. Figure 1.9 gives two examples of carbon nanotubes, respectively, single-walled (Figure 1.9(b)) and multi-walled (Figure 1.9(a)). There is a large range of materials that have been developed in the form of nanotubes: semiconductors [PAR 09], metals [MOR 07, FEN 10], as well as oxides [HU 09].

---

4 Recommendation 2011/696/CE of the Commission, JO L 275 from 20.10.2011

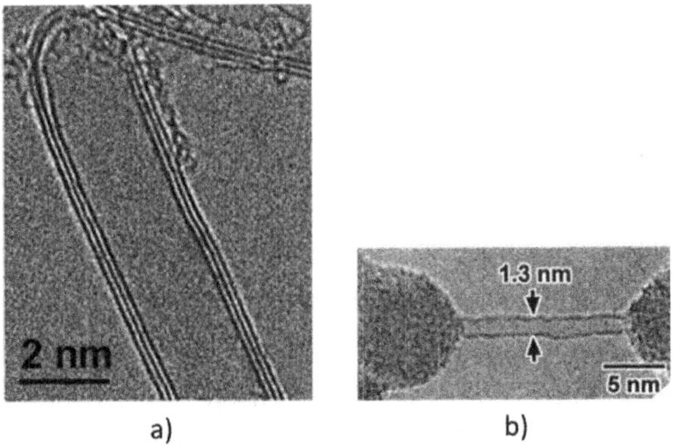

Figure 1.9. *Images obtained by transmission electron microscopy: a) of two multi-walled carbon nanotubes (from [CAM 08]); b) of a single-walled carbon nanotube linked to two silicon carbide electrodes. From [ZHA 99]*

1.2.1.2. *"Full" 1D nano-objects*

It is possible to distinguish here again two large categories: on the one hand nanoribbons, and on the other hand nanowires. However, this distinction is rarely clear and is usually only made indicatively. Nanoribbons possess a section with one side that is much larger than the other (see Figure 1.10(a)), while nanowires, sometimes called nanorods, nanocables, nanowhiskers or nanopillars, have a roughly circular section, even if in reality the section is sometimes polygonal (see Figure1.10(b)). Here as well, it is possible to create "full" 1D nano-objects with many materials: metals [SEL 01], oxides [TSA 05] but especially semiconductors such as GaN [JOH 02], ZnO [HUA 01], GaAs [PER 04], InP [ALG 08], SiGe [POT 11] or SiC [CHO 04] or even nano-objects in which the core does not have the same chemical composition as the exterior of the nanowire: these are called core-shell nanowires (for example, Si–SiO_2 core-shell nanowires [LAU 02] or SiC–SiO_2 core–shell nanowires [ZHA 02]). Nevertheless, the Si nanowires have been the most extensively studied for a long time. In 1964, Wagner and Ellis demonstrated the growth of such nanostructures [WAG 64]. Since then, from the mastering of the growth of Si nanowires [WU 01, SCH 05, SCH 09], to the integration of these nanowires into transistors [HOC 05, LI 06, GOL06], they have been the subject of active research.

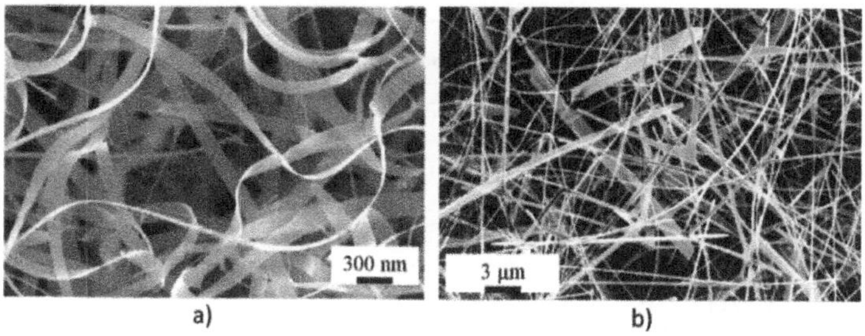

Figure 1.10. *Images obtained by scanning electron microscopy a) of ZnO nanoribbons (from [WEN 03]; b) of SnO_2 nanowires (from [KAC 07])*

1D nanostructures have, therefore, only two dimensions that are smaller than 100 nm, and can be either structurally empty, or structurally full. These 1D nano-objects have been – and still are – the subject of much interest in the research community, especially with regard to semiconducting carbon nanotubes and nanowires in general.

1.2.2. *High surface/volume ratio and its consequences*

The major advantage of nano-objects in general, and 1D nanostructures in particular, comes from the high surface to volume ratio. Indeed, the higher this ratio is, the greater the place occupied by surface atoms. The atoms of a surface distinguish themselves from the atoms located in the volume in several ways. For example, on the surface of a material there are several dangling bonds that influence the other bonds of the atom, as well as the electronic structure. Moreover, the surface does not only affect the first atomic monolayer in direct contact with the empty space, but also the first and second neighboring atoms of the surface. Thus, in a conducting material, electric transport at the surface of a nano-object must be taken into account to better describe the total current. Furthermore, the intrinsic phenomena of interfaces are preponderant at a nanometric scale.

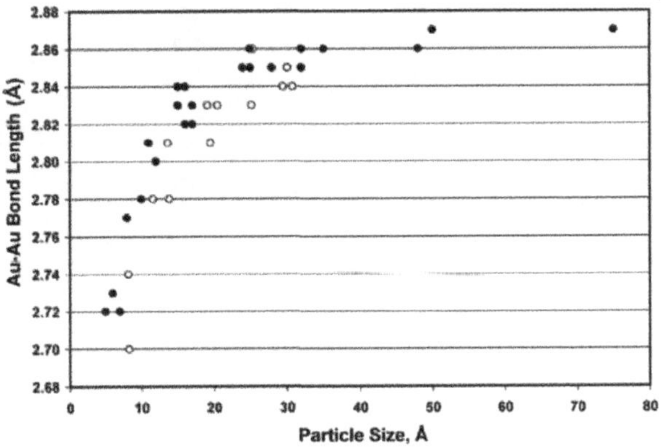

Figure 1.11. *Evolution of the interatomic distance of Au – Au in gold nanoparticles, as a function of the diameter of the particles. From [MIL 06]*

Among the properties that will be affected in this way by the nanometric scale, it is possible to list the following few examples:

– structural properties: the greater the surface/volume, the greater the value of the surface energy, which leads to changes in interatomic distances. If we take the example of gold nanoparticles, Miller *et al.* have shown that the interatomic distance between gold atoms decreases when the diameter of the nanoparticles decreases [MIL 06], as can be seen in Figure 1.11;

– the thermal properties are also affected by the fact that the surface is preponderant. For example, works by Lai *et al.* [LAI 96] have shown that the melting point of tin nanoparticles decreases as the size of the particles decreases, as can be seen in Figure 1.12;

– finally, mechanical properties are also affected by the effects of the nanometric scale. Nickel and copper nanoparticles have a higher critical stress when the diameter of the nanoparticles is smaller (see Figure 1.13);

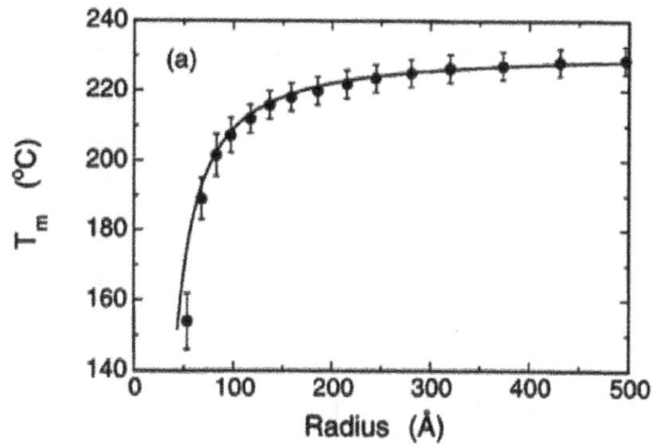

Figure 1.12. *Evolution of the melting point of tin nanoparticles, as a function of the diameter of the particles. From [LAI 96]*

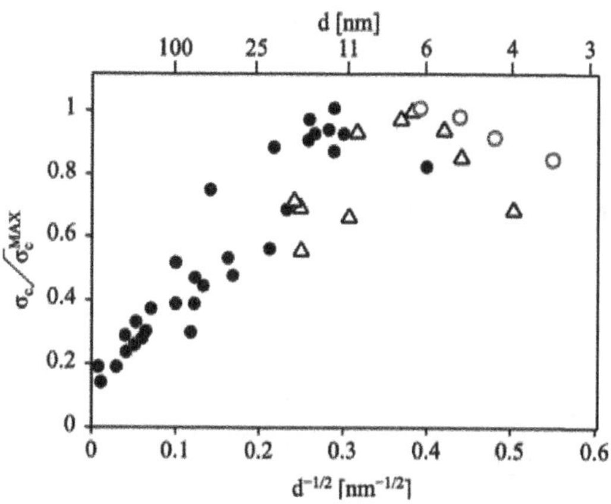

Figure 1.13. *Evolution of the critical stress of nanoparticles of copper and nickel, as a function of the diameter of the particles. From [KUM 03]*

Thus, all the physical properties, as well as the behavior of a material, are impacted when this material is nanometric. This modification comes from the importance of the surface/volume ratio, as well as from the interfaces in relation to the volume.

1.2.3. *Specific properties at the nano metric scale*

Nano-objects, whether one, two or three-dimensional, have been extensively studied, both theoretically and experimentally, not only because of the effects linked to the high surface/volume ratio, but also because of the new properties unique to the nanometric scale. These specific properties come mainly from the modification of the electronic structure of the conduction and valence bands of the materials at a nanometric scale. If we consider the states' densities as a function of the energy of the particles for a bulk material, it appears that covering of the orbitals leads to the existence of conduction and valence bands. These partially cover themselves up in conducting materials, are very farther apart for isolating materials and are close together but separated by a gap for semiconducting materials.

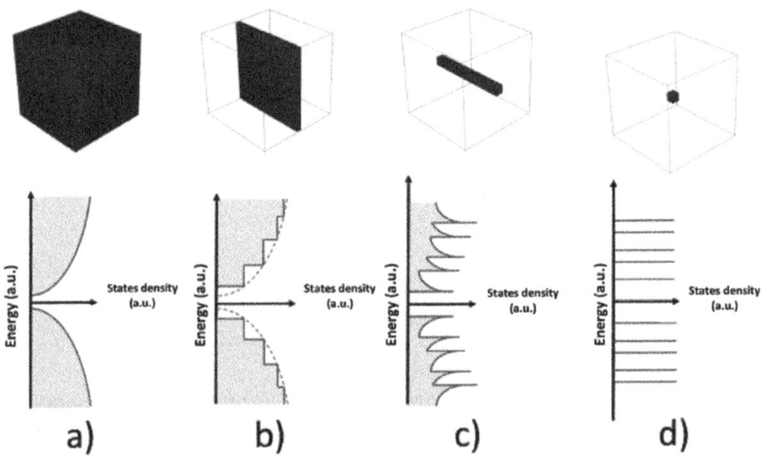

Figure 1.14. *Representation of the states density of objects that a) are volumic, meaning they have no nanometric dimensions; b) have a nanometric dimension: very thin films; c) have two nanometric dimensions: nanowires, nanotubes and nanoribbons and; d) have three nanometric dimensions: nanodots*

However, when the dimensions of an object get close to nanometric sizes, quantum effects, and notably the wavelike behavior of electrons (respectively, the holes) cannot be ignored. Because of this, it appears that the states densities of the electrons (respectively, the holes) are modified and tend toward a quantification of the energy levels possible for the electrons (respectively, the holes). The states densities of the electrons are represented in Figure 1.14, which are the energy levels that are acceptable for the electrons, for objects that have no nanometric dimensions (see Figure 1.14(a)), those with one nanometric dimension (see Figure 1.14(b)), those with two nanometric dimensions (see Figure 1.14(c)) and finally those with three nanometric dimensions (see Figure 1.14(d)). The greater an object's number of nanometric dimensions, the more the energy states of the electrons are quantified, due to the limitation of the movement of the electrons. This is also the case for the holes. This is referred to as quantum confinement in one, two or three dimensions.

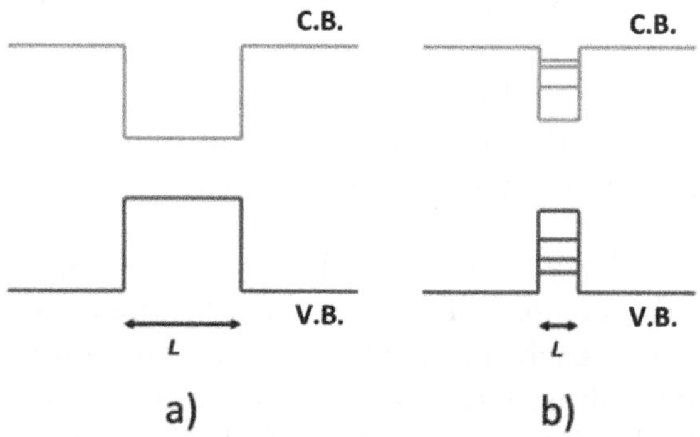

Figure 1.15. *Illustration of a dimension representing a semi-conducting material trapped between two insulators. If length L is small enough, the energy levels in the semiconductor are quantified*

To better grasp the importance of quantum confinement, let us take the example of quantum wells: a semiconductor surrounded by two insulators. Simply put, it is possible to represent (Figure 1.15) the conduction and valence bands for a semiconductor with and without the effects of quantum confinement (Figures 1.15(a) and 1.15(b), respectively). In the first case, the energy necessary to make an electron goes from the valence band to the conduction band is equal to the energy of the gap. However, in the second case, the energy allowed for the electrons (respectively, the holes) being quantified, we first observe an increase of the size of the gap, as well as a quantification of the energies needed to pass from the valence band to the conduction band. This quantification, which comes directly from the resolution of the Schrödinger equation, can be expressed in the following manner:

$$E_n = -E_g + \frac{1}{2m}\left(\frac{n\pi\hbar^2}{L}\right) \qquad [1.1]$$

where E_n represents the energy levels allowed in the conduction band, E_g represents the bandgap energy, taking the top of the valence band as reference, m is the mass of the electron, $\hbar$ is the reduced Planck's constant and finally L, which is defined as the width of the central zone. It is important to note that the dimensions of the quantum wells, and by extension the dimensions of the nano-object, impact directly the effect of quantum confinement.

However, these effects, which are unique to the nanometric dimensions of objects, do not appear for the same dimensions for each material. Indeed, the notion of the exciton Bohr radius, written a_0^*, must be introduced here. It depends on the intrinsic nature of the material, and will allow us to define the limit from which quantum confinement exists. The exciton is defined as the quasi-particle made of an electron–hole pair linked together by Coulomb forces. The radius of this quasi-particle can be expressed in the following manner:

$$a_0^* = \frac{a_0 \varepsilon_r m_e}{m^*} \qquad [1.2]$$

with ε_r being relative permittivity, m_e is the mass of the electron, m^* is the reduced mass that can be written like this:

$$m^* = \frac{m_e^* \times m_h^*}{m_e^* + m_h^*} \quad [1.3]$$

Let us remind ourselves that the Bohr radius a_0, depending on the vacuum permittivity ε_0, the reduced Planck's constant $\hbar$ and e the charge of the electron, among other factors, is equal to:

$$a_0 = \frac{4\pi\varepsilon_0\hbar^2}{m_e e^2} = 52{,}918\, pm \quad [1.4]$$

Let us note finally that the effective masses of the electron m_e^* and of the holes m_h^*, which allow us to take into account the medium in which these particles evolve, can be calculated as a first approximation and by considering that the effective masses are the same in all directions with equation [1.5], where ϵ represents the energy of the electron, which depends on $\vec{k}$, the wave vector[5]. Table 1.4 lists the effective masses of several semiconductors:

$$m_e^* = \hbar^2 \times \left[\frac{d^2\epsilon}{dk^2}\right]^{-1} \quad [1.5]$$

Material	*Si*	*Ge*	*3C–SiC*	*GaAs*	*C (diamond)*
m_e^*	1.08 m_e	0.55 m_e	0.34m_e	0.067m_e	1.87m_e
m_h^*	0.56 m_e	0.37 m_e	0.8m_e	0.45 m_e	0.92m_e

Table 1.4. *Summary of the effective masses of electrons and holes for some semi-conducting materials. From [YOS 95, SON 94, SON 95, KAP 85, IOF 13, SAU 04]*

5 We can also talk of dispersion relation for equation $\epsilon = f(\vec{k})$.

Thus, when an object possesses at least 1D smaller than a_0^*, a quantum confinement appears. This value depends on the material through the effective masses of the electrons and of the holes m_h^* and of the relative permittivity ε_r. In the case of 1D nano-objects, the diameter would, therefore, need to be smaller than a_0^* to be able to obtain quantum confinement. Figure 1.16 represents the value of the exciton Bohr radius for different semiconductors as a function of the bandgap. It appears that few materials possess an exciton Bohr radius greater than 10 nm (see Figure 1.16).

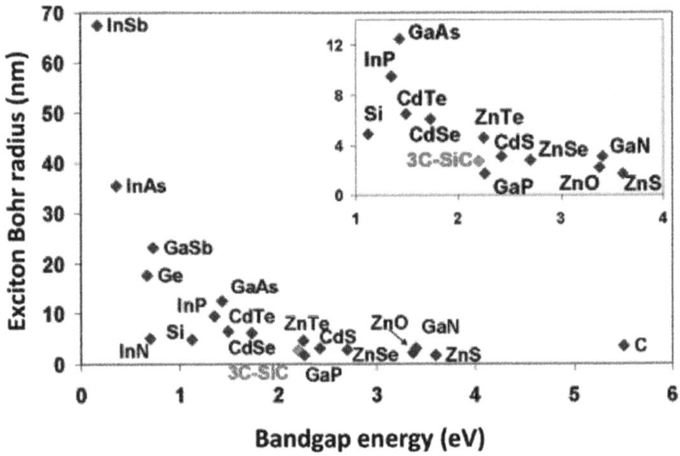

Figure 1.16. *Representation of the value of the exciton Bohr radiusas a function of the bandgap for different semiconductors. From [WU05, YOF 01, NIQ 00, OST 05]*

The physical properties unique to nanometric scales, therefore, depend on the nature of the material used. While we refer to them as nano-objects as soon as 1D is smaller than 100 nm, it is clear that quantum effects only appear for dimensions that are much smaller, in the order of 10 nm. This separation between nanometric objects and the properties specific to the nanometric scale allow us to define two classes of nano-objects: those whose dimensions allow the appearance of quantum effects and those whose dimensions do not. These two classes of nano-objects have not been developed in the same manner: while quantum objects are still in the early stages of fundamental research, larger nano-objects are already being used in industry. From this remark, it must be noted that 3C–SiC is not one of the most interesting materials in terms of the effects of quantum confinement, due to small value of the exciton Bohr radius of this material ($\approx$ 2.7 nm [WU 05]). We will see soon that such dimensions are hard to reach even today, which explain why applications targeted by 1D 3C–SiC nanostructures do not use the properties unique to the nanometric scale. The main advantage of these objects currently resides in the obtaining of large surface/volume ratios so as to benefit from greater sensitivity.

1.3. Conclusion

The physical and chemical properties intrinsic to SiC reflect the advantages of developing such a material. Among its remarkable properties, we can list the following:

– the thermal properties of SiC, notably its high operating temperature and the excellent thermal conductivity of SiC, which is better than that of copper;

– the electronic properties of SiC such as the strong breakdown field and the good saturation velocity of the carriers;

– the resistance of SiC in harsh environments: high temperatures, acid and basic media, and oxidizing media;

– SiC is the only semiconductor that is both biocompatible and hemocompatible;

– the polytypism of SiC limits its development to grow zero-defect single crystals. The cubic form of SiC is the only one that has isotropic properties and an indirect bandgap energy less than the other polytypes.

The stakes present in the growth of 1D SiC nanostructures are, therefore, to ally these physical and chemical properties with the effect of nanometric dimensions. As emphasized previously, quantum effects are hard to reach in SiC nano-objects; however, the effects caused by the high surface/volume ratio are promising and warrant the studying of 1D SiC nanostructures.

2

State of the Art of the Growth of SiC-1D Nanostructures

The exhaustive description of all the methods used in the growth of SiC-based one-dimensional (1D) nanostructures present in the scientific literature makes it hard to get an overall vision; the goal of this chapter is therefore to offer a large panorama of the different processes for the synthesis of silicon carbide (SiC) based 1D nanostructures. From an experimental point of view, three types of nanostructures appear: SiC nanowires, SiC nanotubes and finally core–shell nanowires.

2.1. State of the art of the growth of SiC nanowires

In this part we look at the three main methods for the synthesis of SiC nanowires by the bottom-up approach, i.e. the silicidation of carbon nanotubes, which was the first used historically, growth by the Vapor–Liquid–Solid (VLS) mechanism, and finally a derivative of the VLS mechanism, elaboration in the gaseous phase.

There is a large diversity of techniques used for growth by the bottom-up approach, and this for one simple reason: the chemical formula for SiC contains the two atoms that constitute the most studied nano-objects: silicon nanowires and carbon nanotubes. Similarly, from a purely chemical point of view, the equation that leads to the formation of SiC (equation [2.1]) is very simple and the sources of silicon and carbon atoms are well-known. From this simple chemical reaction we can list three growth processes that will be developed next, in function of the nature of the sources of carbon and silicon atoms: either the source of carbon is solid and the source of silicon is gaseous, or the two sources are gaseous, or else either the source of silicon is solid and the source of carbon is gaseous (see section 2.1.5).

$$\mathrm{Si} + \mathrm{C} \rightarrow \mathrm{SiC} \qquad [2.1]$$

2.1.1. *Silicidation of carbon nanotubes*

Historically, the first 1D SiC nano-objects were obtained through a process where the source of silicon is gaseous and the source of carbon is solid. Dai *et al.* demonstrated in 1995 the growth of the first SiC nanowires by the reaction of a silicon oxide [SiO] gas with carbon nanotubes at high temperatures [DAI 95]. This reaction, called silicidation, allows the use of carbon nanotubes as a matrix, thus controlling the size of the SiC nano-objects formed. However, the crystalline quality of the SiC obtained is mediocre (see Figure 2.1). Moreover, it is not easy to obtain SiC nanowires in this way because of the carbon nanotube's internal empty space, as shown by Wallis *et al.* [WAL 08]; with the same experiment, but because of the dispersion of diameters of the carbon nanotubes, it is possible to obtain SiC nanowires (see Figure 2.2(b)) and SiC nanotubes (see Figure 2.2(a)). As the source of silicon should be in the form of gas, either it is directly injected as a gas into the growth apparatus, for example in the form of tetramethylsilane (TMS) [MO 02], or else it is formed *in situ* by the passage of a SiO powder [PAN 00, SUN 02] (or even an SiO_2 powder [KEL 03]) into the gaseous phase at high temperature. A vector gas, argon or dihydrogen, allows the transport of the silicon atoms in the gaseous phase to the carbon nanotubes.

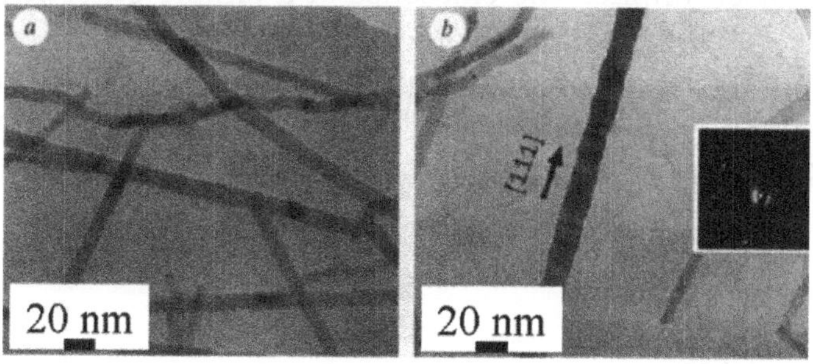

Figure 2.1. *TEM images of the first SiC nanowires, obtained through a mechanism that causes the carbon nanotubes to react with a gaseous source of silicon. From [DAI 95]*

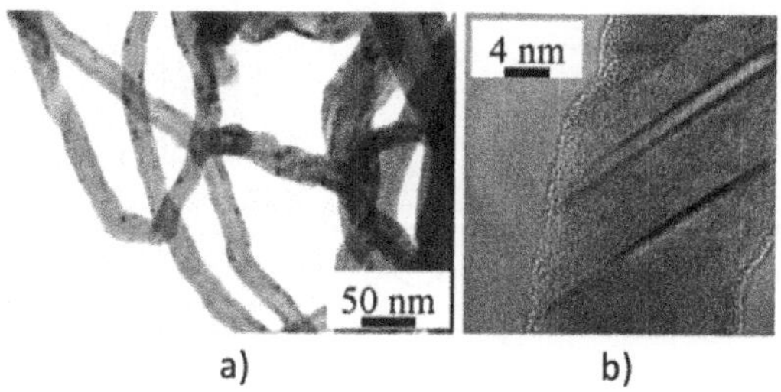

Figure 2.2. *a) TEM image of SiC nanotubes, obtained through the reaction of an SiO gas on the carbon nanotubes. b) TEM image of a SiC nanowire obtained through the reaction of an SiO gas on the carbon nanotubes during the same experiment. From [WAL 08]*

2.1.2. *Synthesis through the VLS mechanism*

The VLS growth mechanism was discovered in 1964 by Wagner and Ellis [WAG 64] for silicon microwires. The VLS mechanism calls upon a liquid element into the growth of nanostructures, usually metallic, which acts as a growth vector. This is referred to erroneously as a catalyst, despite the fact that, strictly chemically speaking, the presence of this element does not increase the rate of the reaction[1]. The choice of catalyst will be the least miscible element possible at low temperatures in the material being deposited, and which often implies the existence of a simple eutectic in the binary phase diagram of the catalyst-material to be deposited. So as to better understand the VLS growth mechanism, we will describe the example of SiC nanowires, using iron as the catalyst (see Figure 2.3).

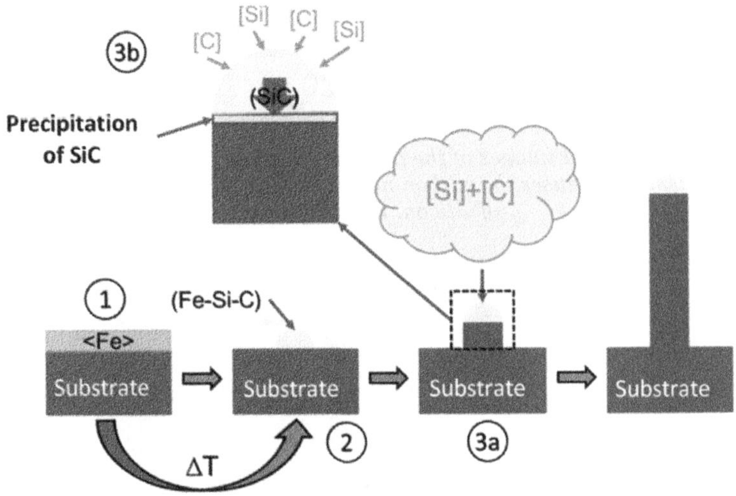

Figure 2.3. *Schematic representation of the growth of SiC nanowires by the VLS mechanism: the catalyst, here iron deposited by evaporation, forming a thin layer (1), is dewetting at the surface by increasing the temperature that determines the diameter and position of the future nanowire (2); next, a gaseous source containing both carbon and silicon atoms cracks at the surface of the droplet until it saturates it (3a), leading to the precipitation of the silicon carbide under the droplet (3b) and causing the nanowire to grow*

Droplets of the catalyst are deposited on a substrate where the nanostructures will grow (either these droplets are deposited directly in the form of colloids, or they are

1 A catalyst is defined as an element favoring a chemical reaction, and that comes out of it unchanged.

obtained by dewetting at the surface of the substrate (1)). The droplets thus formed will determine the diameter of the nanostructures, as well as their position. Let us note that the nanostructures will be in an epitaxial relation with the substrate used, which provides this technique with a definite advantage.

The system is then brought to a temperature at which the Fe–Si–C alloy becomes liquid (2). Once this step has been carried out, a precursor gas for the element to be deposited is placed in contact with the Fe–Si–C droplets (3a). This precursor gas – in excess at this point – will crack at the surface of the catalyst, which is supersaturated (3b), and precipitates the material underneath the droplet, leading to the growth of the nanowires. The name given to this growth mechanism can therefore be explained by the presence of a vapor phase (the precursor gas), a liquid phase (the liquid alloy formed with the catalyst) and a solid phase (the nanowire).

There are many reasons that explain the interest that exists for this technique. First of all, the localization and control of the dimensions of nanowires are the main advantages; also, such a growth is comparable to a chemical vapor deposition (CVD), a tried, tested and industrialized technique. The nanostructures obtained in this way are of good crystalline quality due to the intermediate growth in liquid phase. This growth mechanism also has a large number of parameters that can be changed, for example doping the material by adding a gas during growth, or changing the material during growth to obtain heterojunctions.

However, it is important to note that the VLS mechanism becomes a lot more complex in the case of the elaboration of SiC nanowires. First of all, the choice of the catalyst is problematic, since the solubilities of Si and C need to be equivalent for this catalyst, with a melting point adapted to the growth of SiC: at high temperatures, the risk of losing matter during growth is high. The second problematic point involves the gaseous phase. Indeed, even if several strategies can be envisaged to obtain a gaseous phase that is rich in both Si and C atoms (introduction of two different precursor gases, one for the carbon, the other for the silicon; use of single precursor gas containing Si and C, use of a solid source, etc.), its stability remains an issue.

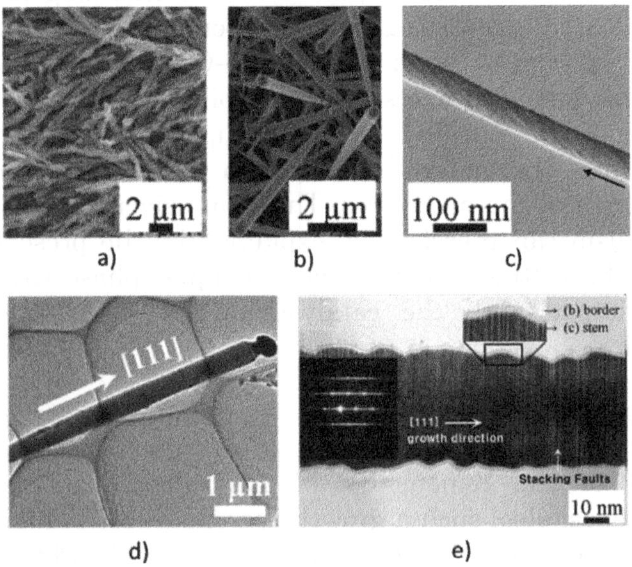

Figure 2.4. *a) SEM image of SiC nanowires obtained by using diethylmethylsilane as a precursor gas for the carbon and the silicon, and nickel as the catalyst, (from [HYU 09]); b) SEM image of SiC nanowires obtained by VLS with polysilazane as the source of carbon and silicon atoms, with iron as the catalyst (from [WAN 08]); c) TEM image of a SiC nanowire obtained by VLS with carbon chloride as the source of carbon atoms and a Si substrate as a source of the silicon; the catalyst used is nickel (from [ATT 09]); d) TEM image of a SiC nanowire by VLS using polysilazane as the source of Si and C atoms (from [FEN 12]); e) TEM image of a SiC nanowire obtained by VLS. We can see several stacking faults along the growth axis (from [KAN 03])*

However, the VLS mechanism has been used in the growth of SiC nanowires. Generally, the catalyst retained is nickel [HYU 09, ATT 09], although other materials have also been used, such as iron [WAN 08], palladium or platinum [SUN 07] or even magnesium [XI 06]. The precursors of the Si and C chemicals are diverse; however, due to the high growth temperature of SiC, these are usually cracked into simple molecules in the gaseous phase [CHA 06]: Si, SiC, SiC_2, C or even the different silicon hydrides (SiH_x), carbon hydrides (C_xH_y). The SiC nanowires obtained in this way have a variety of morphologies, as can be seen in Figure 2.4. Most of the SiC nanowires have a morphology that reveals problems with the control of growth: indeed, a variation of the diameter along the nanowire is reflective either of a change of temperature, even a very slight one, during growth, or of a constant loss of the atoms of the catalyst, leading to a decrease of diameter of nanowire. This leads to numerous structural defects, and mainly a high density of stacking faults, as can be seen in Figure 2.4(e).

2.1.3. *Development in the gaseous phase – VS mechanism*

The Vapor–Solid (VS) mechanism is the easiest mechanism for obtaining 1D nanostructures. It consists of evaporating the material that we wish to use and condensing it into the desired form in another location. Seeing that 1D nanostructures are stable structures, it is usually sufficient to apply a thermal gradient between the evaporation zone and the condensation zone to be able to obtain these structures. An important point that needs to be considered however is the evaporation of the material. Depending on the nature of the source of the atoms (gas, liquid, solid; compounds to be deposited pure, or alloys; pure reagents or mixtures), the evaporation method changes. The main problem with this development technique is the difficulty in controlling the positioning of the nanostructures, as well as in controlling the steps of the growth of the material. Because of this, the nanostructures are often of mediocre crystalline quality.

Figure 2.5. *a) TEM image of a SiC nanowire coated with carbon obtained using the VS mechanism (from [BEC 05]); b) SEM image of SiC nanowires obtained using the VS mechanism (from [ZHA 09a]); c) TEM image of SiC nanowires obtained using the VS mechanisms, the structural defects leading to the formation of curvatures in the wires (from [YAN 05b]); d) TEM image of a SiC nanowire obtained using the VS mechanism where we can observe a twin and stacking faults (from [NHU 02])*

This is the case of SiC nanowires obtained using this mechanism. The sources of silicon and carbon can be either provided by a same molecule, TMS for example [NHU 02], either by two different sources: C_3H_6 gas and SiO powder [LI 03], silicon and polypropylene powder as presented in Figure 2.5(a) [BEC 05, BEC 07a] or even SiO and carbon powder obtained from the graphite crucible used [ZHA 09a]. The nanowires obtained are usually long (see Figure 2.5(b)) but many crystalline defects are present in SiC, as can be seen in the TEM image of Figure 2.5(d), where a twin can be seen, as well as stacking faults.

2.1.4. *Carburization of Si nanowires*

This process is similar to the VS mechanism, except that here, the silicon remains in the solid form, and the carbon atoms are supplied through the gas: this is the carburization of silicon nanowires. This technique is also called carbonization of silicon nanowires. While the carburization of silicon for the epitaxy of SiC on Si has been well-studied (see section 3.2), the application of this technique to nanowires has not been much explored. Despite this, the technique possesses interesting advantages: the absence of a catalyst during the carburization phase limits contamination caused by high temperatures, the crystalline quality of the silicon nanowires offers a good starting point for the SiC nanowires, and finally doping is mostly controlled in the case of silicon nanowires.

The first team to publish on this topic is that of Zhang *et al.* [ZHA 00]. While their goal was first of all to deposit carbon nanotubes on the silicon nanowires, SiC nanowires were grown, as shown in the TEM image of Figure 2.6(a). Although this is interesting, the result was not exploited further, since the article itself was seeking to demonstrate the possibility of depositing carbon nanotubes on the surface of the silicon nanowires.

However, Figure 2.6(b), which presents a TEM image of SiC nanowire obtained by Tsakalakos *et al.* [TSA 05] through the carburization of silicon nanowires in the CVD growth apparatus from flows of carbonaceous gases such as CH_4 or C_3H_8, is more interesting. Even if this result is part of the annex of a more complex work on the carburization of molybdenum nanowires [CHE 12], it is the most accomplished: using silicon nanowires obtained *in situ* through the VLS mechanism, the carburization mechanism is very close to those carried out for epitaxial growth of SiC on Si. Nevertheless, the crystalline quality of the SiC nanowires obtained is not excellent and the edges of the nanowires are damaged.

Finally, in 2008 Yang *et al.* presented another experiment of silicon nanowires carburization [YAN 08]. Starting with the silicon nanowires obtained by anisotropic etching of a silicon substrate, the carburization is carried out in an apparatus brought to 1350°C and a flow of ethanol carried by argon serves as a carbon precursor. The result is presented in the form of a SEM image in Figure 2.6(c). The SiC nanowires obtained in this way are amorphous and present a highly porous morphology.

Figure 2.6. *a) TEM image of an SiC nanowire obtained by carburization of a silicon nanowire using a carbon nanotube growth apparatus (from [ZHA 00]); b) TEM image of an SiC nanowire obtained by carburization of a Si nanowire using C_3H_4 (from [TSA 06]); c) SEM image of SiC nanowires obtained by carburization of Si nanowires with ethanol (from [YAN 08])*

We will see in Chapter 3 that it is possible to greatly improve the crystalline quality of the SiC obtained through the carburization of Si nanowires, at this process appears promising for 1D nanostructures other than SiC nanowires.

2.1.5. *Conclusion on the growth of SiC nanowires*

SiC nanowires usually crystallize in the cubic form[2], due to the low growth temperature (around 1000°C) for SiC [KNI 68] and contain a very high planar structural defects density, such as stacking faults and twins (see Figure 2.4). These characteristics are shared by all the methods of the growth of SiC nanowires and constitute a barrier to the crystalline improvement of the material that is hard to overcome. The creation energy of these defects is very low [HON 00] due to the thermodynamic metastability of the cubic SiC (compared to the hexagonal form), and would even

2 4H- and 6H–SiC nanowires were also obtained by choosing hexagonal SiC seed monocrystals and/or by doping with aluminum to stabilize the hexagonal form [WAN 10, KRI 11, WEI 08]; however, stacking faults are still observed inside these nanowires.

be negative for the creation of a stacking fault in 3C-SiC [LIN 03].

2.2. State of the art of the growth of SiC nanotubes

As for carbon nanotubes, SiC nanotubes are the objects of many theoretical studies (*ab-initio* simulations) that envisage this structured material's potential as a gas sensor [ZHA 09b, WU 11], chemical sensor [CHE 12, GAN 11] or in the storage of hydrogen [MPO 06, BAN 11]. While the development of new applications that use C nanotubes is currently limited by the mix of metallic and semiconducting CNTs, the possible uses for SiC nanotubes do appear to be very promising[3].

However, the growth of SiC NTs remains difficult, and as a result there are few experimental works present in the literature on this subject [LAT 14]. Most syntheses use the silicidation of carbon nanotubes, presented as a shape-memory synthesis technique. However, the morphological and crystalline qualities that result are often problematic, with the appearance of defects and of porosity in the walls [NHU 02, LAT 14]. The internal and external diameters of the nanotubes are also no longer controlled and a mix of nanowires and nanotubes is usually obtained following the silicidation of carbon nanotubes [SUN 02, TAG 05]. We must mention that an interesting experimental work was carried out by Sun *et al.* on the reaction of carbon nanotubes with gaseous silicon monoxide (from the sublimation of a SiO powder) [SUN 02]. Within a mix of different 1D nanostructures, multi-walled SiC nanotubes have been highlighted (Figure 2.7(b)). The allotropic transformation under an electron beam of the

3 It is potentially possible to control the semiconductor or metallic character of carbon nanotubes by modifying their surface chemistry [SOU 13].

3C-SiC nanotubes has, however, stopped the full characterization of the crystalline structure of these original nanotubes.

The synthesis of SiC nanotubes has also been carried out by the selective chemical etching of ZnS–SiC core–shell nanowires [HU 04]. These nanotubes are polycrystalline and present a variety of morphologies (straight, curved or zig-zag). To the best of our knowledge, only the work by Pei *et al.* in 2006 [PEI 06], by condensation of SiC and SiO powders at high pressure – 8 MPa, has resulted in the spontaneous growth of SiC NTs without having to use a *template* material, even if some Si nanotubes are supposed to form at the beginning of this synthesis reaction. To illustrate this, most of the SiC nanotubes synthesized so far have been presented in Figure 2.7.

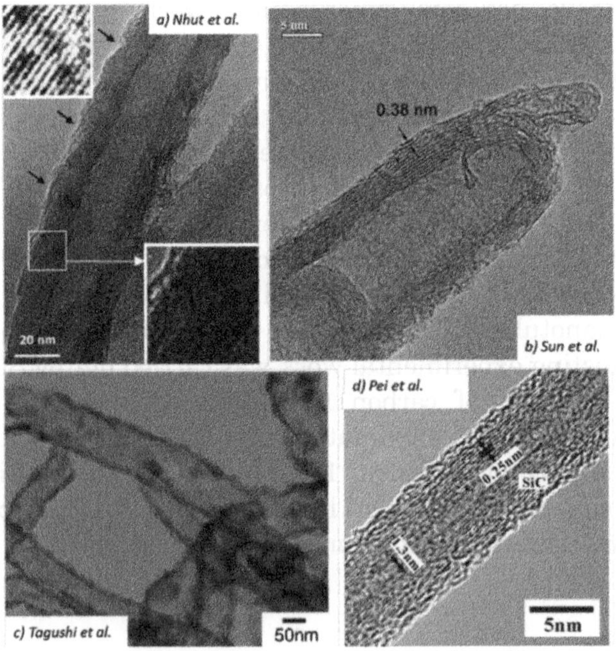

Figure 2.7. *Morphology of the SiC nanotubes – TEM images – synthesized in the state of the art, from [SUN 02, NHU 02, TAG 05, PEI 06]*

2.3. State of the art of the growth of SiC-based core–shell nanowires

2.3.1. *Si–SiC core–shell nanowires*

In the literature, to our knowledge and despite the advantages that these objects present (notably with regards to the protective aspect of the SiC shell), only five research teams have demonstrated obtaining Si–SiC core–shell nanostructures, which we describe here very briefly (see Figure 2.8) [LAT 14].

Si–SiC microspheres have been developed by thermal plasma carburization of Si powder, of micronic granulometry [LI 07]; in this case, the layer of SiC has been formed by carburization of liquid Si. This layer of SiC is polycrystalline, and empty spaces have been characterized at the Si–SiC interface.

One technique of depositing by ion beam was used by Zhou *et al.* [ZHO 00] to coat Si nanowires through the reaction of methane with silicon. The structural characterizations carried out by TEM revealed that 3C–SiC nanocrystals partially covered the surface of the Si nanowires. This surface modification led to the observation of a photoluminescence spectrum whose efficiency is tripled.

Nanocomposites of Si–SiC have also been obtained through a two-step process [BEA 11]. The Si nanowires (called *nanotowers* here, due to their small size) are obtained through the well-known VLS mechanism, catalyzed by gold, followed by a deposit of polycrystalline 3C–SiC with a thickness of 250–350 nm obtained using a plasma technique. The particularly rough SiC shell suggests a mediocre crystalline quality. In this study, the SiC shell was created so as to generate compressive stress within the core of Si to study plasticity.

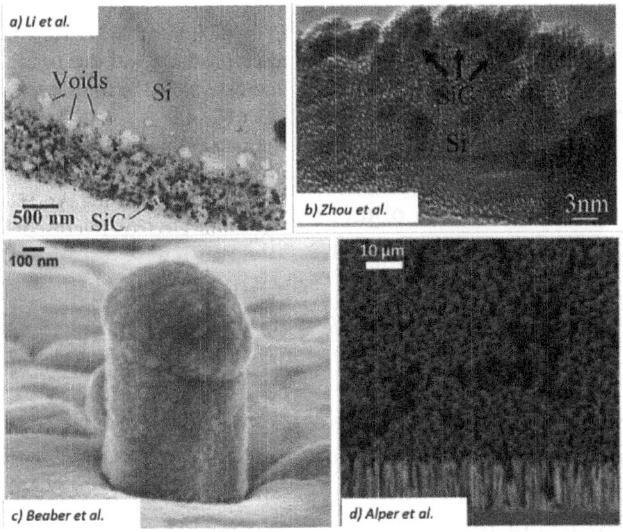

Figure 2.8. *Si–SiC shell–core nanostructures reported in the literature. a) From Li et al., b) from Zhou et al., c) from Beaber et al., d) from Alper et al. From [LAT 14]*

Finally, Si nanowires were coated with SiC so as to create a robust electrode material for super-capacitors [ALP 12]. The Si nanowires were synthesized by wet etching, while the SiC was deposited by CVD. The crystalline quality of the material was not studied, but a low level of crystallinity is expected due to the particularly rough morphology of the system formed. The work shows that this material could be used in future high performance super-capacitors (see section 4.3).

2.3.2. *Other SiC-based core–shell nanowires*

Despite the interest in the core-shell design, there is relatively little mention of SiC-based core-shell nanowire systems in the relevant literature.

Nanowires with a SiC core and a silica shell remain the most common, since a simple oxidation of the SiC nanowires allows us to obtain a passivating SiO_2 shell. Some teams have managed to directly obtain a coaxial SiC-SiO_2 structure (via a carbonated or silicide precursor, already oxidized [MEN 07, ATT 08]), like in the CVD reaction of carbon monoxide with the Si substrate, catalyzed by nickel [ATT 08]. It has also been shown that shells made of amorphous silicon could increase the optical emission of the core–shell system [FAB 10].

With the goal of reinforcing composite matrices, SiC–C core–shell nanowires were synthesized [YAN 05a, KIM 03]. The additional deposit of the amorphous carbon on the SiC nanowires was grown to adjust the interfacial energy between the ceramic matrix and the reinforcement fibers.

Other core–shell nanowires have been developed in the last few years: NiSi–SiC [NAZ 14], SiC–AlSiC [FAN 12] and SiC–BN [LI 05, BEC 07b], from 2005 to 2014. One of the advantages of a BN shell is to isolate the semiconductor SiC core, while ensuring excellent chemical inertia.

2.4. Conclusion

In the literature, most of the works on the growth of 1D SiC-based nanostructures revolves around the growth of SiC nanowires. There are several methods to synthesize SiC nanowires, all based on the bottom-up approach (favoring the “growth” aspects), as opposed to the top-down approach (favoring the “etching” aspects). We have listed three main processes for SiC nanowires:

– direct growth through VLS or VS (Vapor–Solid) alignment with;

– the silicidation of C nanotubes;

– the carburization of Si nanowires.

These different techniques lead to the synthesis of cubic SiC nanowires, with a high density of planar structural defects, such as stacking faults. The silicidation of C nanotubes also allows the obtaining of SiC nanotubes with nanometric wall thicknesses. SiC-based core–shell nanowires are less common, and, in the case of Si–SiC nanowires, the SiC shell remains polycrystalline.

3

An Original Growth Process: The Carburization of Si Nanowires

Whichever the growth mechanism, the crystalline quality of the silicon carbide (SiC) nanowires (NWs) always seems to be less than satisfactory; as a result, we have chosen to work on the carburization of Si NWs, still mostly unknown, whose advantages include the excellent starting point that are single-crystalline Si NWs of controlled size, coupled with knowledge on the carburization of Si substrates. The study of the carburization of Si NWs has led to the development of original structures: core–shell Si–SiC NWs and SiC nanotubes. The experimental application of the process is first presented here, followed by silicon carburization in the context of the epitaxy of thin layers of 3C–SiC. The two nanostructures obtained through the carburization of Si NWs are described, as well as the diagram of the existence of these Si-SiC core-shell NWs and nanotubes. The mechanisms of carburization at the origin of one-dimensional (1D) nanostructures are discussed in depth.

3.1. Si nanowires

Let us remind ourselves that two approaches can be considered in developing Si NWs: a bottom-up approach (growth) or a top-down approach (etching).

Bottom-up approach: the bottom-up approach consists of assembling the NW, atoms at a time, while the top-down approach consists of removing matter, again atoms at a time, resulting in an NW. The Si NWs are typically obtained using the bottom-up approach, via growth in a silane vapor phase, catalyzed by gold [LAT 08], calling on the vapour–liquid–solid (VLS) mechanism explained previously (see section 2.1.2).

Top-down approach: the use of plasma etching (top-down approach) also results in Si NWs [MAR 13]. We shall not reiterate the elementary mechanisms that govern plasmas; however, we can say that a plasma is made up of charged particles such as ions and electrons, and of chargeless particles, referred to as neutral. The processes of plasma etching call upon two types of etching initiated by the species present in the plasma:

– chemical etching, which is carried out by neutral species, and is therefore isotropic and selective;

– physical etching, which corresponds to an ionic bombardment, due to the polarization of the substrate to be etched. This process is highly anisotropic and not very selective.

Plasma-assisted etching allows us to get the best out of these two processes, being both selective and anisotropic. Hereafter, we describe in a simplified manner a process of the plasma etching of Si NWs from a silicon substrate oriented along the axis [100] (see Figure 3.1):

1) First of all, we must define the patterns that will be transferred into the silicon during etching, and for this a

mask must be prepared. The choice of mask is particularly important in the case of the etching of vertical NWs, since the mask must resist as best possible to the etching plasma, so that the NWs have a strong aspect ratio. For this reason, it is necessary to use a hard mask, for example, which is made up of a layer of silicon dioxide (SiO_2) deposited by chemical vapor deposition (CVD) (see Figure 3.1(a)).

2) A photoresist is then spin coated and exposed to deep ultraviolet (DUV) photolithography (248 nm) to define the patterns (see Figure 3.1(b)).

3) To transfer the pattern previously inscribed into the photoresist into the hard SiO_2 mask, a first plasma etching is carried out using, for example, a plasma made up of the gases $CF_4/CH_2F_2/N_2$, a highly selective mixture, which will etch the silicon dioxide, leaving the remaining photoresist intact (see Figure 3.1(c)).

4) The etching of the NWs by plasma itself can then be carried out, taking care nonetheless to remove the layer of photoresist beforehand (oxygen plasma, chemical treatment, etc.). The choice of the mixture of gas, as well as its proportions, is obviously crucial in order to carry out an etching of high quality. The $SF_6/O_2/HBr/SiF_4$ mixture is an example of a plasma composition that results in the Si NWs presented in Figure 3.3:

i) the SF_6 and SiF_4 species containing fluoride atoms are known for their ability to etch silicon. However, they cause considerable lateral etching;

ii) the O_2 species causes the generation of a passivation layer of the Si_xF_y type, which limits lateral etching. Moreover, this species limits the etching of the hard SiO_2 mask;

iii) boron hydride HBr also leads to the formation of a redeposit of the SiO_xBr_y type on the sidewalls (see Figure 3.1(d)).

5) The last step of this process consists of removing the hard mask by chemical attack using hydrofluoric acid in the case of a SiO_2 mask (see Figure 3.1(e)).

Let us note that ideally, the length of the SiNWs etched depends directly on the time of exposition to the plasma. However, this etching is limited by the mask's resistance to the plasma.

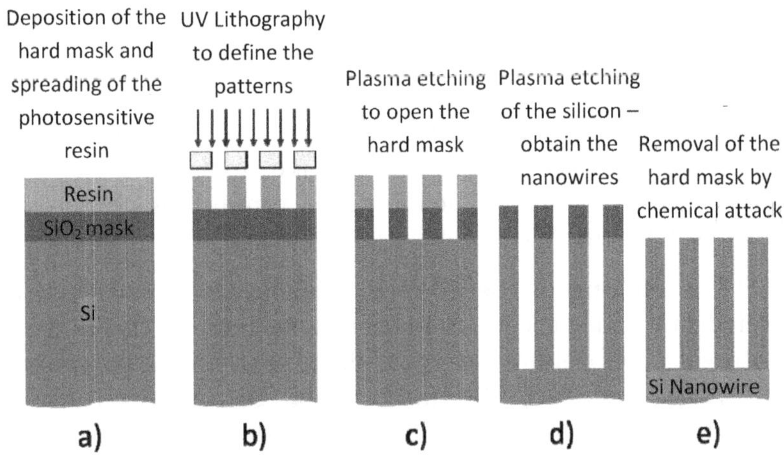

Figure 3.1. *Schematic representation of the different stages of the etching of silicon nanowires. a) Deposition of the SiO_2, to be used as the hard mask for plasma-assisted etching, followed by spin coating of the photoresist used as the soft mask for the transfer of the pattern to the hard mask. b) UV photolithography of the pattern made up of blocks of a certain diameter chosen for the nanowires. c) Hard mask etching assisted by plasma. This plasma is made up of the gaseous mixture of $CF_4/CH_2F_2/N_2$. d) Plasma-assisted etching of the silicon nanowires. The plasma is made up of the gaseous mixture $SF_6/O_2/HBr/SiF_4$. e) Removal of the hard SiO_2 mask by hydrofluoric acid chemical attack. From [MAR 13]*

Comparison of the bottom-up and top-down approaches: whichever the technique chosen, the Si NWs are single crystalline. However, while the size (notably the diameter), the orientation and the density of the NWs are variable when the NWs are grown by VLS, the Si NWs obtained by etching are calibrated by size and ordered at the surface of the silicon (see Figure 3.2).

Figure 3.2. *SEM images of Si nanowires obtained using the bottom-up approach on the left or using the top-down approach on the right (from [OLL 13])*

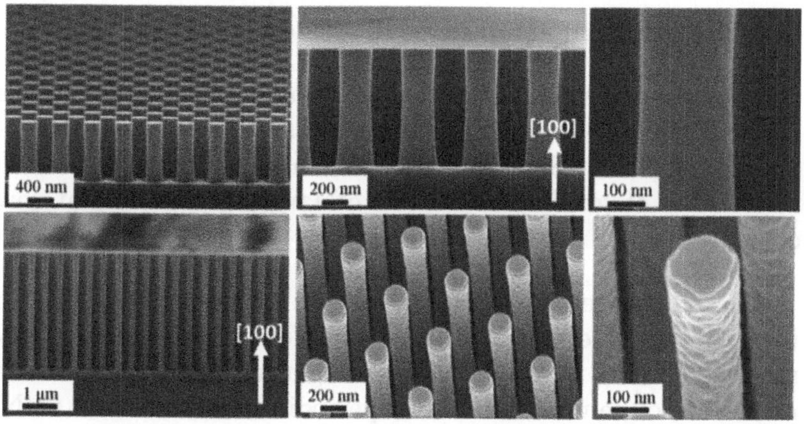

Figure 3.3. *SEM image of the Si nanowires, obtained using the top-down approach, used in the study of carburization. Short Si nanowires (top) and long Si nanowires (bottom)*

The carburization of Si NWs obtained using the top-down approach benefits from known parameters (diameter, identical crystalline direction of the NWs and organization of the nanostructures). This choice also has the advantage of avoiding the delicate step of the elimination of the gold catalyst (placed at the top or on the sidewall of the Si grown by VLS [DEN 08]): this chemical cleaning is not perfect, and traces of gold usually persist afterwards.

The sidewalls of the Si NWs obtained by etching can be slightly rough when the NWs are long (see Figure 3.3). This is due to the heterogeneity of the passivation layer formed during etching. We shall see hereafter that this roughness can be eliminated through an appropriate rapid thermal annealing (see section 3.4.2.5). In this study, two types of Si NWs obtained by etching, all oriented in the crystalline direction [100], were used: short NWs with a length of 1 µm and a diameter of 250 nm; and long NWs with a length of 5 µm and a diameter of 200 nm. The figure shows the morphology of these Si NWs.

3.2. The carburization of bulk silicon

The carburization of bulk silicon was vastly studied at the start of the 1980s in support of the studies on the hetero-epitaxy of SiC on Si. These studies aimed to obtain cubic SiC substrates of good quality and of large dimension, at low cost.

Lastly, the hetero-epitaxy of SiC on Si has experienced an increase in interest in the production of layers of III-N materials. This acronym groups together all of the nitride materials of column III of the periodic table, such as GaN, BN, AlN and InN. Apart from their good electronic characteristics, it is the direct "gap" of some of these materials that has spurred their development, with the goal of producing Light-Emitting Diode. However, it is difficult to obtain substrates that are big enough for these materials. The advantage of SiC relies here on a correct match of the lattice constants with the III-N materials (see Table 3.1). For example, the wurtzitic forms of AlN and GaN have lattice constants close to those of hexagonal polytypes 4H–SiC and 6H–SiC. Currently, the SiC substrates are already used in the epitaxial growth of layers of AlN and GaN, despite the small diameters of the SiC substrates. The development of the hetero-epitaxy of SiC on Si acts as an intermediate step before the growth of the III-N material on greater SiC surfaces.

Material	Lattice constant *a* and *b* (Å)	Lattice constant *c* (Å)	Space group
3C–SiC	4.359	–	$F\bar{4}3m$
2H–SiC	3.081	10.061	$P6_3mc$
4H–SiC	3.081	15.08	$P6_3mc$
6H–SiC	3.073	15.08	$P6_3mc$
15R–SiC	3.073	37.70	$P6_3mc$
GaN (wurtzite)	3.189	5.186	$P6_3mc$
GaN (zinc-blende)	4.52	–	$F\bar{4}3m$
AlN (wurtzite)	3.11	4.98	$P6_3mc$
AlN (zinc-blende)	4.38	–	$F\bar{4}3m$
BN (wurtzite)	2.55	4.17	$P6_3mc$
BN (zinc-blende)	3.615	–	$F\bar{4}3m$
BN (hexagonal)	2.5–2.9	6.66	$P6_3mmc$
InN (wurtzite)	3.53–3.54	5.69–5.70	$P6_3mc$
InN (zinc-blende)	4.95–4.98	–	$F\bar{4}3m$

Table 3.1. *Crystallographic data on the different SiC polytypes, as well as on the main III-N materials. From [STR 92, BAG 09, MOR 09]*

However, the hetero-epitaxy of SiC on Si encountered two problems that are intrinsic to the materials used:

– the lattice mismatch between Si and SiC. Indeed, the lattice constant of silicon, which crystallizes in a cubic system of the diamond type, is of $a = 5.43$ Å. It results in a lattice mismatch of 19.7% between 3C–SiC and silicon. Moreover, the hexagonal SiC polytypes require growth temperatures greater than the melting temperature of silicon. Thus, it appears that only the cubic polytype of SiC can be envisaged in the case of the hetero-epitaxy of Si on Si. The difference of lattice constant between Si and 3C–SiC can also be expressed in the following manner: at the SiC/Si interface, one bond out of five is not satisfied;

– the difference of thermal expansion coefficient between the two materials is contained between 8 and 11% at ambient temperature: Si = 2.6×10^{-6} K^{-1}, according to [OKA 84]; 3C–SiC = 2.77×10^{-6} K^{-1} according to [SLA 75]. While this value may seem quite low, this gap becomes preponderant at high temperatures of epitaxial growth, which then leads to cooling of the constraints at the SiC/Si interface, and to the appearance of structural defects.

Direct epitaxy of 3C–SiC on Si through a classical technique of deposition in a vapor phase leads to the production of an unusable 3C–SiC layer, as it is too constrained, and therefore subject to fracture and it is also polycrystalline. This problem was overcome due to work by Nishino *et al.*, who in the 1980s added a carburization step to the silicon substrate, just before the epitaxial growth of 3C–SiC [NIS 83]. This carburization, sometimes called carbonization, consists of reacting a precursor carbon gas with the surface of the silicon so as to obtain a very thin layer of SiC, also called buffer layer, which serves as a seed to the direct growth of 3C–SiC. Nishino *et al.* thus proposed the epitaxial growth of SiC on Si in three steps, which still serve as a base today:

– the first step consists of cleaning the silicon substrate and notably deoxidizing it using a flow of H_2 (dihydrogen) at high temperatures. Let us remark that the pressure is usually fixed between 10^{-2} Pa and the atmospheric pressure. Typically, the substrate is kept for 5 min at 1000°C under the dihydrogen flow;

– the second step allows the formation of a buffer layer of SiC by carburization due to a CH_4 methane flow or a C_3H_8 propane flow. The carburization dwell is usually higher than the deoxidation dwell, 1150°C for example, and lasts 5 to 10 min. Let us note here that currently, an additional step is added just after carburization. This step consists of leaving the atoms time to diffuse, still at high temperature, but without a precursor gas;

– finally, the expitaxial growth of SiC is carried out. On top of the flow of a carbon precursor gas, the flow of a silicon atom precursor gas is used. This growth is all the more efficient if carried out at high temperatures, or closer to the melting point of silicon.

The carburization step being crucial in obtaining a layer of good-quality SiC, it is therefore the subject of much research. It is important to understand the specificities of this carburization step to be able to cover the mechanisms that take place during carburization. Indeed, as opposed to classical epitaxial growth, here, the substrate reacts with the precursor gas. It is actually for this reason that we use the term of reactive CVD to better describe this step. As a result, the interface between the layer of SiC formed and the substrate evolves during deposition since the latter serves as a source of silicon for the growth of the SiC layer.

The questions of the Si/SiC interface and of the growth mechanisms of the SiC layer were discussed even before the works by the S. Nishino team. The conversion of the silicon substrate into SiC substrate by carburization was considered as a production process, although the crystalline quality of the SiC formed was not good. Graul and Wagner [GRA 72] investigated the carburization of silicon to obtain a polycrystalline layer of SiC. By using radioactive tracers, they observed that the formation of the buffer layer of SiC was accompanied by the diffusion of silicon atoms toward the surface of the SiC layer. Mogab and Leamy observed the formation of empty structures at the Si/SiC interface, which they attributed to a growth mechanism of the SiC layer by diffusion of the Si atoms at the surface of the grains [MOG 74]. An example of these empty structures, also called "etch-pits", is presented in Figure 3.4. However, it was only in the middle of the 1990s, with the works of Li and Steckl [STE 92, LI 95], that a reliable model became available for the carburization of silicon as well as the formation of etch-pits, despite the fact that diffusion had already been identified by Graul and Wagner as being the parameter limiting carburization [GRA 72]. The model provided by these researchers relies on one phenomenon in particular: out-diffusion of silicon through the SiC. As it turns out, the diffusion of Si atoms in SiC is quicker than the diffusion of C atoms in SiC: in Figure 3.5, the diffusion coefficients are reported for the Si and C atoms in SiC as a function of the temperature. We can also note the data-grouping work by Cimalla *et al.* [CIM 00], who placed on a same graph the measured values of the diffusion coefficient of Si atoms through the SiC layer obtained during the carburization step by different teams using different growth methods (see Figure 3.6).

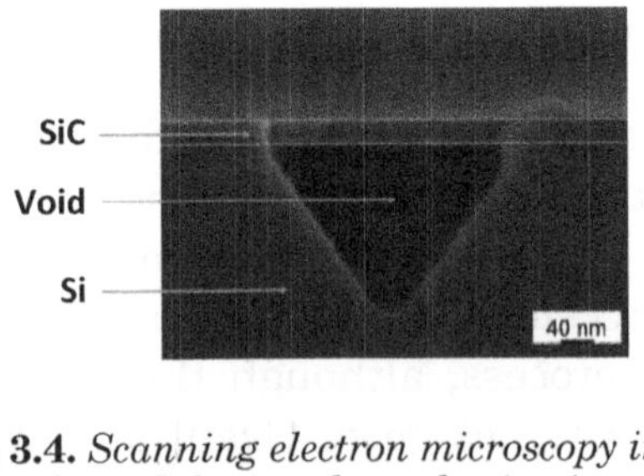

Figure 3.4. *Scanning electron microscopy image of an etch-pit formed during the carburization of silicon. The SiC layer is visible above the void. From [OLL 13]*

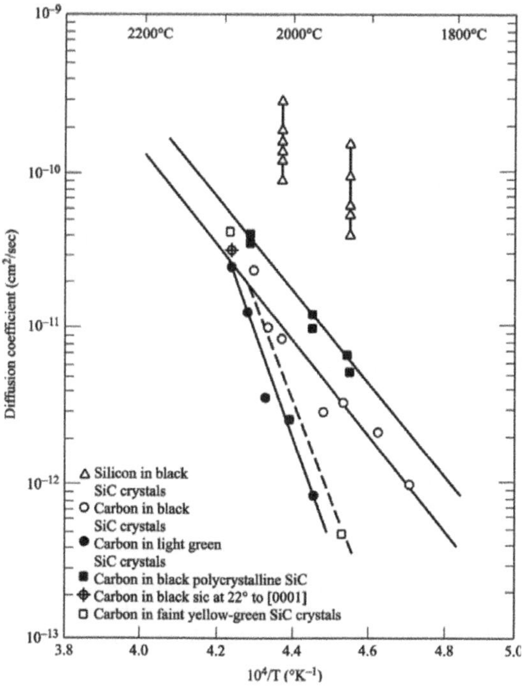

Figure 3.5. *Value of the diffusion coefficients of Si and C atoms in SiC as a function of the temperature; measurements made at 10⁴ Pa. From [GHO 66]*

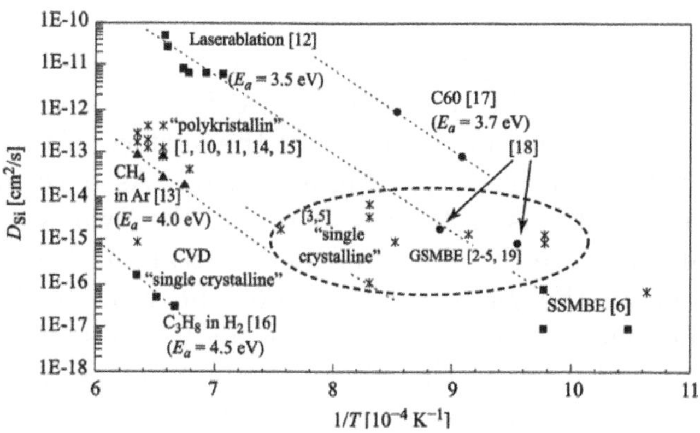

Figure 3.6. *Value of the diffusion coefficients of Si atoms in SiC during the carburization step as a function of the temperature for different experiments. From [CIM 00]*

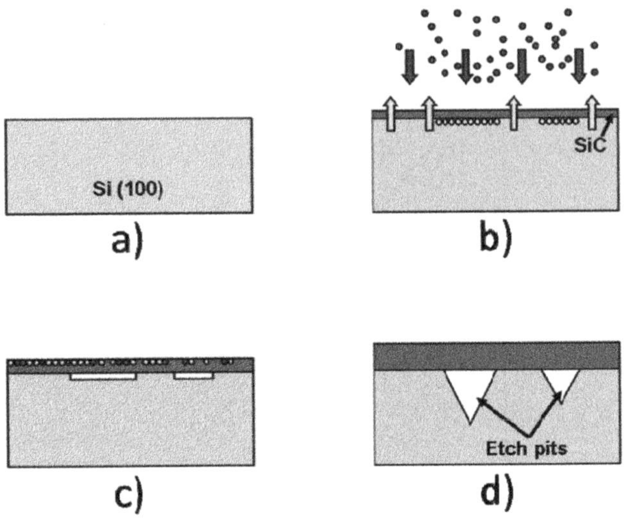

Figure 3.7. *Diagram representing the phenomenon of out-diffusion taking place during the carburization of silicon. a) Si substrate oriented along the [100] axis. b) The precursor gas of C (in dark gray) reacts with the outer layer of Si to form a first SiC atomic monolayer. c) The Si atoms (light gray) out-diffuse through this SiC atomic monolayer and react with the carbon atoms present at the surface due to the low diffusion velocity of C compared to the Si atoms in SiC. (c) Voids are formed under the SiC layer due to diffusion, d) thus revealing the {111} dense planes*

From this point, it is possible to explain the mechanism of out-diffusion that appears during the carburization of Si. Starting with a silicon substrate (see Figure 3.7(a)), a first SiC atomic monolayer is formed at the surface of the silicon and covers it entirely (see Figure 3.7(b)). As a result, this monolayer acts as a barrier between the two sources of reagents that are the silicon of the substrate and the carbon contained in the gaseous phase. For the growth of SiC to continue, one of the two reagents needs to cross this barrier (see Figure 3.7(b)). As the silicon atoms diffuse quicker than the carbon atoms in SiC, it is the substrate atoms that are directed toward the surface and that react to form SiC. This loss of Si atoms in the substrate then results in the formation of etch-pits under the SiC (see Figure 3.7(c)). Due to the crystallographic structure of silicon, oriented along the [001] axis, in this example these etch-pits present a pyramidal structure, which show the {111} dense planes (see Figure 3.7(d)).

The out-diffusion of silicon through SiC during carburization explains why it is difficult to obtain a layer of SiC of good quality by carburization alone. The key is therefore to find the right parameters to obtain a SiC buffer layer that is thick enough and of good enough crystalline quality to act as a seed for the epitaxy of SiC [FER 96a, FER 96b], while limiting the formation of pyramidal structures in silicon, since these weaken the whole of the structure.

3.3. Experimental application

3.3.1. *Carburization apparatus*

The carburization apparatus is made up of a horizontal, tubular furnace, heated resistively between 20 and 1,200 equipped with a pressure-regulating system. The pressure can be controlled between 0.2 and 10^5 Pa. A gas input/output system is connected to the furnace to control atmospheric composition. Six different gases can be used: carbon monoxide, carbon dioxide, argon, dihydrogen, propane and methane. The reachable flow rate ranges for the three gases that we used during this study are the following:

– for Ar: 0–200 $mL.min^{-1}$;

– for H_2: 0–200 $mL.min^{-1}$;

– for CH_4: 0–4 $mL.min^{-1}$.

A diagram illustrating the carburization apparatus is presented in Figure 3.8.

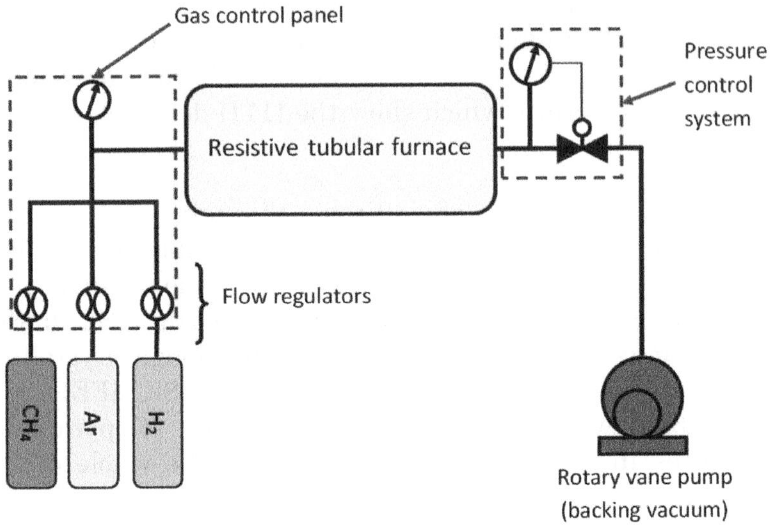

Figure 3.8. *Simplified schematic representation of the carburization apparatus*

3.3.2. *Methods of characterization*

As for any material study, the appropriate methods of characterization must be chosen. In this work, scanning electron microscopy has been used routinely to observe the morphology of 1D nanostructures. Raman spectroscopy, which can provide the nature of the polytype and determine the presence of possible faults nearly instantaneously, as well as provide qualitative information on the crystalline quality, is only used rarely. This is also the case for X-ray diffraction, as the shell (or the layer) of SiC obtained following carburization is usually too thin to be detected. Using Raman spectroscopy, from a thickness of the SiC of approximately 100 nm on a Si substrate, the signature of the SiC can be observed (see section 3.5, Figure 3.30(b)).

The structural characterizations can be carried out using electron microscopy. Double focused ion beam–scanning electron microscope FIB-SEM has largely been used in the realization of different thin lamella for observation under transmission electron microscopy TEM, *in situ* in FIB-SEM (at 30 keV), or *ex-situ* by TEM (at 200 keV).

Three types of thin lamella are prepared (see Figure 3.9). These are:

– a longitudinal cross-section on a planar layer, the Si substrate;

– a longitudinal cross-section through the 1D nanostructure;

– or a transverse cross-section through the 1D nanostructure (equivalent to a NWs "washer").

The use of these microscopies allows us to access the structural (crystalline structure and orientation of the crystalline planes) and physical/chemical (chemical composition, notably) characterization of the objects synthesized to determine *in fine* the mechanisms involved during carburization.

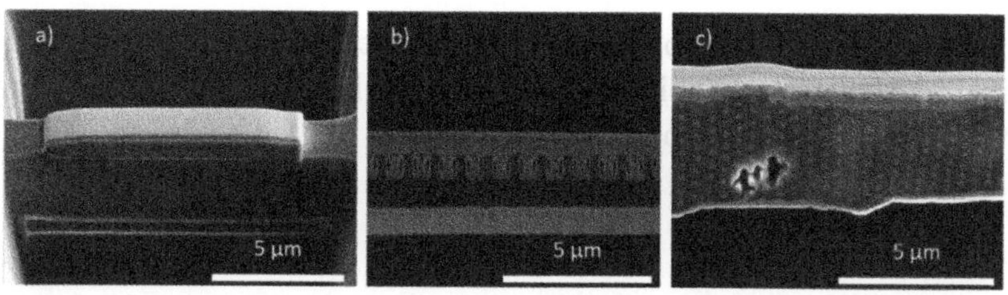

Figure 3.9. *SEM images of the three types of thin lamella: a) longitudinal section on a planar layer, b) longitudinal section through the 1D nanostructures and(c) transverse section through the 1D nanostructures. A deposition of C or Pt – visible at the top of each lamella – is done in-situ in FIB-SEM to avoid any damage (such as amorphization) to the surfaces of the sample*

3.4. Growth of core–shell Si–SiC nanowires

3.4.1. *Introduction*

This particular nanostructure benefits of two advantages:

– a Si core for which doping is easily controlled via the etching of the Si substrate, and which ensures good electronic transport;

– a SiC shell whose properties of chemical inertia and of biocompatibility[1] could be used advantageously.

One objective, notably, could be the use of these core–shell Si–SiC NWs in the production of nano-sensors, functioning in harsh or biological environments.

The first experimental study method used to validate this approach therefore consists of creating a NW transistor-type device, for which the SiC surface is functionalized upon the grafting of deoxyribose nucleic acid (DNA). A variation of the conduction must be detected when the strands of DNA are grafted. The recognition of a positively (or negatively) charged species leads to a depletion (or accumulation) zone, and therefore a drop (or increase) in the conductivity in real time.

This study has focused on the synthesis of the material, its associated physical and chemical properties and the understanding of growth mechanisms.

---

1 The association of properties of chemical inertia with the possibility of chemical functionalization can seem contradictory *a priori*. In reality, silicon carbide is not attacked by basic and acidic solutions at ambient temperature, which is an indicator of its chemical inertia. But it can also be functionalized chemically via steps of hydroxylation and of silanization to create bridging oxygen at the surface of the silicon carbide. It is the chemical inertia of SiC that ensures its biocompatibility.

3.4.2. *Experimental study*

3.4.2.1. *Preliminary study*

The preliminary works consist of preserving the integrity of the Si NWs during the increase in temperature before carburization, which happens at temperatures approximately 1000°C under dihydrogen. The presence of H_2 is necessary for the reduction of the native oxide present in the Si NWs. This study can appear unremarkable, but in reality, it is accompanied by experimental work that is of consequence.

If the increase in temperature happens at low-pressure levels (approximately 12 Pa), the NWs become very damaged. From 600°C, the surface of the sidewalls of the NWs is modified, and the sublimation of Si becomes considerable at 1100°C (Figure 3.10). If the pressure is increased up to atmospheric pressure, the NWs are intact up to 900°C, in accordance with the Le Chatelier principle[2] (Figure 3.11). The sublimation of Si is also influenced by the presence of residual oxygen. If we compare the morphology of the Si NWs after a temperature increase up to 800°C at 12 Pa, the NWs are damaged in the furnace where the tube is made up of quartz, while the NWs remain intact when the furnace's quartz tube is covered in silicon (Figure 3.12).

The sublimation of Si is logically favored at lower pressure levels (approximately 10 Pa), and worsened by the presence of oxygen. These results are supported by thermodynamic considerations [LAT 11]. The thermodynamic calculations, whose principle is based on the minimization of free enthalpy, are carried out due to the simulation software Facstage™, with the *Solid and Gases Pure Substances* (SGPS) database.

2 "When the external changes brought to a system in equilibrium lead to a new state of equilibrium, the system opposes these changes and counteracts their effects", statement by Prince d'Henry Le Chatelier, 1884.

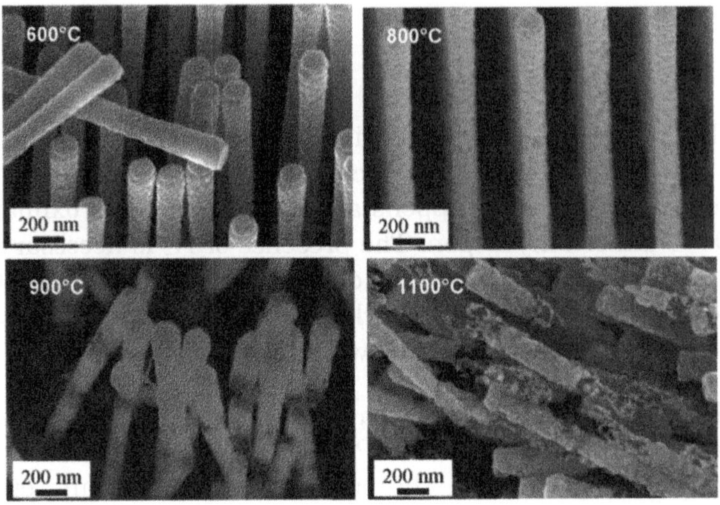

Figure 3.10. *SEM images of Si nanowires having undergone an increase in temperature under a flow of H_2 at a pressure of 12 Pa at 600, 800, 900 and 1100°C*

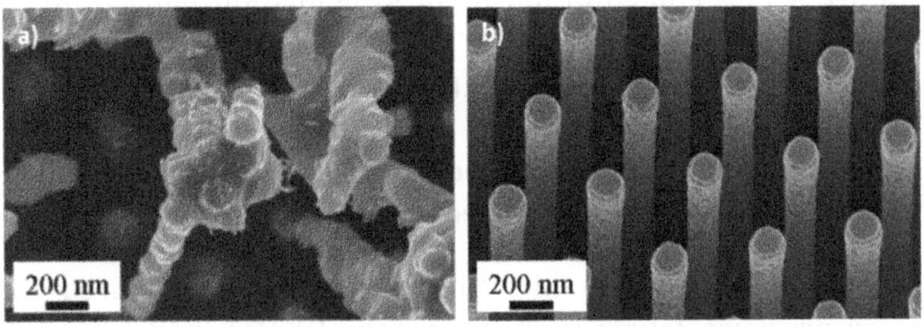

Figure 3.11. *SEM images of Si nanowires having undergone an increase in temperature under a flow of H_2 at 900°C and for a total pressure of: a) 13 and b) 10^5 Pa*

These calculations show that at 1200°C, the system of solid silicon and gaseous dihydrogen is relatively stable whatever the pressure. However, the addition of the dioxygen into the gaseous phase consumes all of the silicon by forming volatile SiO species at low pressure (10 Pa) (see Table 3.2).

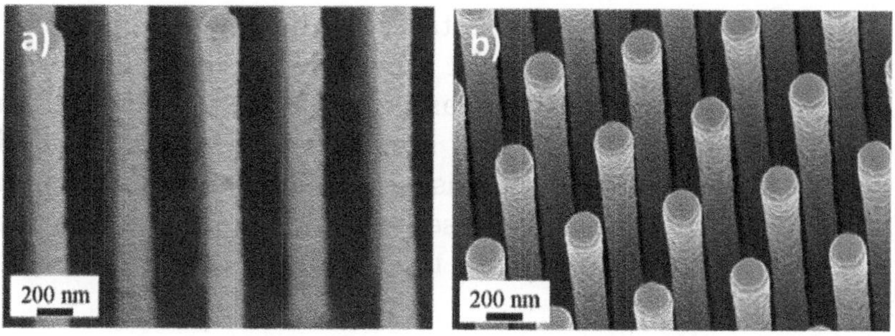

Figure 3.12. *SEM images of the long Si nanowires having undergone an increase in temperature at 800°C under a flow of H_2 at a pressure of 12 Pa: a) in a bare quartz furnace and b) in a silicon-covered quartz furnace*

Initial system	⟨Si⟩	[H_2]	⟨Si⟩	[H_2]	[O_2]
Initial quantities (moles)	2	1	20	10	1
Final system (1200°C)	⟨Si⟩	[H_2]	⟨Si⟩	[H_2]	Si in the vapor phase especially [SiO]
Final quantities (moles) at P = 10 Pa	2×10^{-4}	1×10^{-3}	18	10×10^{-3}	2
Finally quantities (moles) at $P = 10^5$ Pa	2×10^{-5}	1×10^{-4}	19.98	10×10^{-4}	0.02

Table 3.2. *Comparison of the two simulated systems ⟨Si⟩ / [H_2] and ⟨Si⟩ / [H_2] / [O_2] at a temperature of 1200°C for a pressure of 10 Pa and a pressure of 10^5 Pa*

To keep the Si NWs intact during the temperature increase, we choose to work at atmospheric pressure so as to keep the species in a solid phase, and we start carburization at 800°C so as to form a first seed layer of SiC that should provide more protection against possible sublimation than a Si surface. To reduce the presence of oxygen in the gaseous phase, the quartz used in the carburization furnace is replaced by alumina.

3.4.2.1.1. Experimental condition for the synthesis of core–shell Si–SiC nanowires

For all these carburization experiments, the pressure has therefore been fixed at the level of atmospheric pressure. During the temperature rise under a mix of H_2 and Ar, the methane used as a precursor gas is injected, starting at 800°C with a flow rate of 0.5 mL.min^{-1}. When the temperature reaches the dwell, which varies between 1000 and 1200°C depending on the experiment, the methane flow is increased up to 4 mL.min^{-1}. The dwell time, written as t_2, also varies between 5 and 60 min. All of these experimental conditions are represented in Figure 3.13.

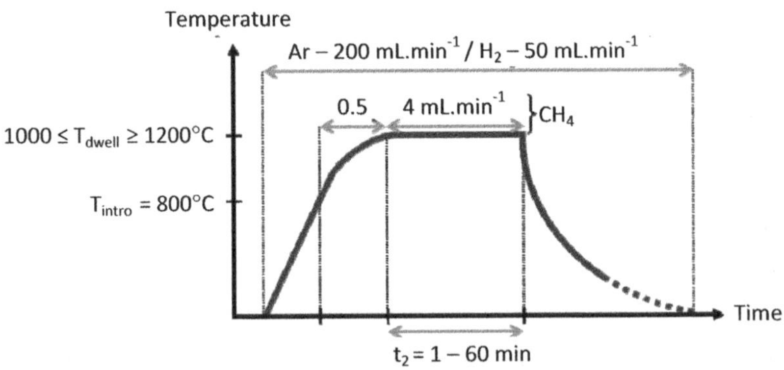

Figure 3.13. *Schematic representation of the thermal profile and the gas flows used in the growth of core–shell Si–SiC nanowires*

3.4.2.2. *Choice of carburization temperature*

The carburization temperature varies between 1000 and 1200°C for carburization times fixed at 20 minutes. Transverse thin lamellas are prepared on the planar part of the (100) Si substrates. At 1000°C, the thickness of the SiC is extremely thin and a layer cannot be measured using scanning TEM (STEM) images obtained *in situ* in the FIB-SEM at 30 keV. At 1100°C, a thin layer of SiC is obtained at the surface of the Si (with a thickness of approximately 3 nm). At 1150°C, the morphology of the wires becomes rough, and the roughness becomes greater at 1200°C. This roughness is caused by the presence of C, as can be seen in the STEM images (see the contrast difference between the Si/SiC/C). Raman spectroscopy can confirm the presence of amorphous carbon with the appearance of C and G bands centered approximately 1,350 and 1,600 cm^{-1}. However, Raman spectroscopy cannot here determine the presence of SiC (too thin). The TEM observations of thin lamellas have, however, shown without a doubt the presence of 3C–SiC on the silicon (100) (Figure 3.14).

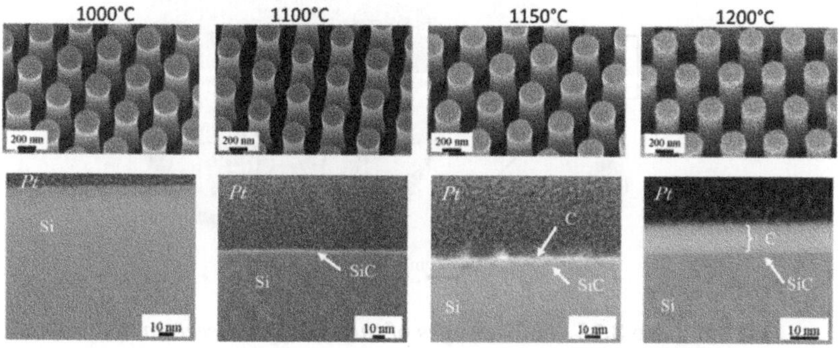

Figure 3.14. *(Above) SEM images of Si–SiC nanowires. (Below) STEM images at 30 keV of the longitudinal planar sections for dwell temperatures of 1000, 1100, 1150 and 1200°C. The carburization time at the dwell remains constant and equal to 20 min*

Under these experimental conditions (chosen methane/flow), 1100°C seems to be the optimal temperature for the carburization of Si without any amorphous carbon deposition (for a fixed dilution in the dihydrogen).

But is this SiC deposit, carried out here on a substrate of Si (100), transposable to NWs coming from the etching of such a substrate?

To answer this question, a transverse thin lamella is prepared on long carburized NWs at 1100°C, and then observed by TEM (Figure 3.15). This observation reveals a uniform deposit of single-crystalline cubic SiC all around the Si with a thickness of 2.8 nm (Figure 3.16), which is identical to the thickness measured for the planar layer in a high-resolution TEM image (Figure 3.17). Let us note that the TEM observation of the section of the NW also shows an artificial thickness of approximately 20 nm due to the initial roughness present on the long Si NWs.

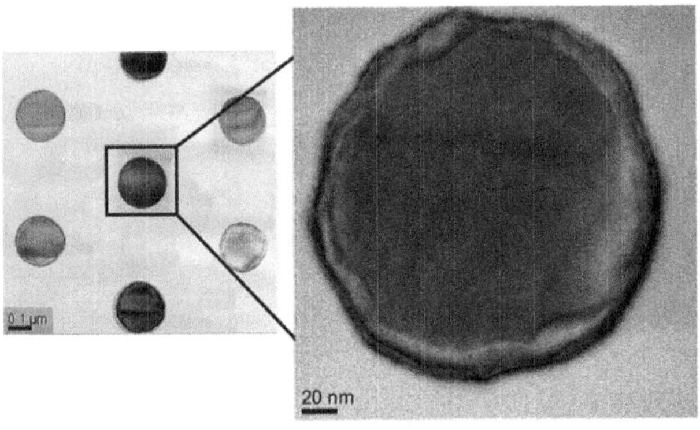

Figure 3.15. *TEM images of the transverse section of core–shell Si–SiC nanowires carburized at 1100°C*

The SiC shell therefore has an equivalent thickness no matter the crystalline surfaces, and it appears to be single crystalline from the TEM observations. However, due to the large difference between the lattice constant of SiC and the lattice constant of silicon, which is of 20%, the presence of dislocations seems to be inevitable [BHA 98].

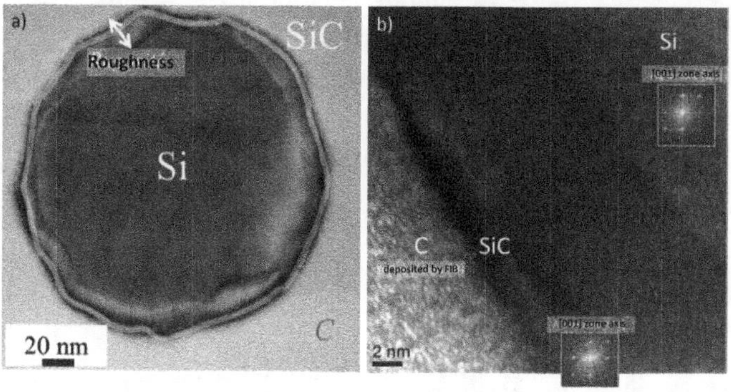

Figure 3.16. *a) TEM image of the transverse section of Si–SiC NW with decoration of the SiC shell and b) high-resolution TEM image of the SiC shell present on the nanowire observed along a [001] zone axis for a carburization temperature of 1100°C, with the Fourier transforms of the core and the shell in the boxes*

SEM images of long and short Si NWs, before and after carburization, are presented in Figure 3.18 and show that under these carburization conditions, the morphology of the NWs remains unchanged.

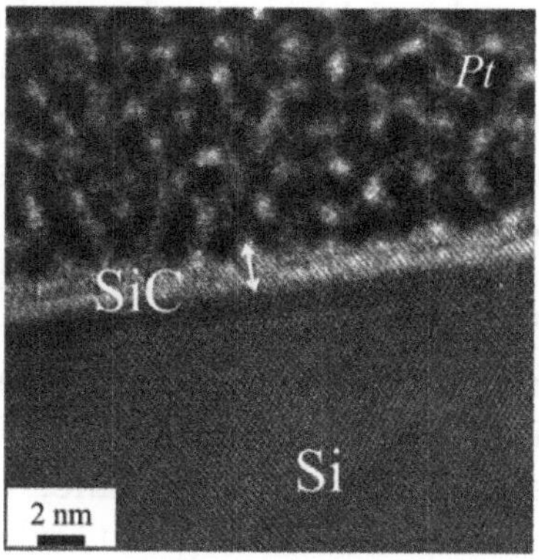

Figure 3.17. *HR-TEM image of a carburized planar layer at 1100°C observed along a longitudinal section*

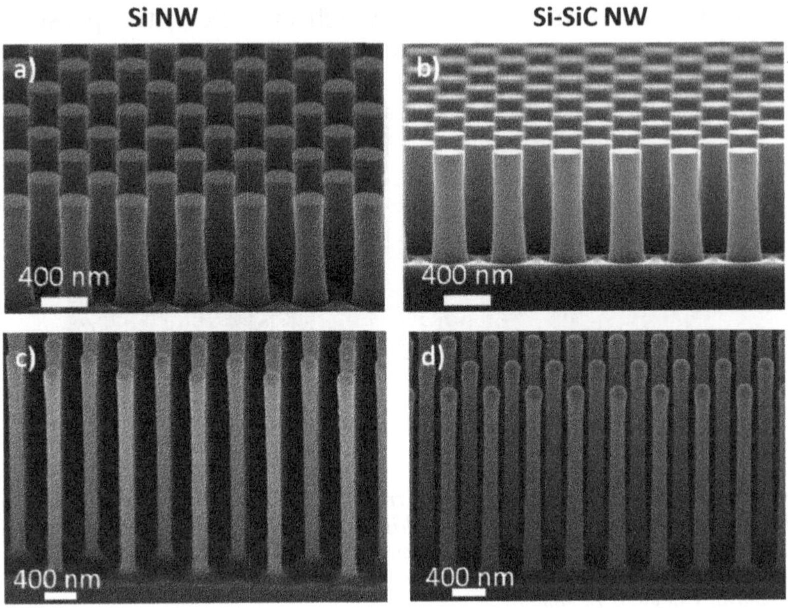

Figure 3.18. *SEM images a) of short Si nanowires, c) of long Si nanowires, b) of short Si–SiC nanowires and d) of long Si–SiC nanowires*

3.4.2.3. *Conformity of the SiC shell*

The microscopic characterization have shown that the SiC shell is formed in an isotropic manner and that the deposit is consistent (same thickness), however it still needs to be verified whether this deposit is continuous over the whole length of the NWs. A simple way of founding this out is by carrying out a chemical etching on the Si NWs and the carburized Si NWs. After attack by potassium hydroxide at room temperature, the Si NWs are completely destroyed, but the core–shell Si–SiC NWs remain intact (Figure 3.19). This result, which is macroscopic this time, is the sign of a consistent and continuous SiC deposit, which is primordial before considering any further applications.

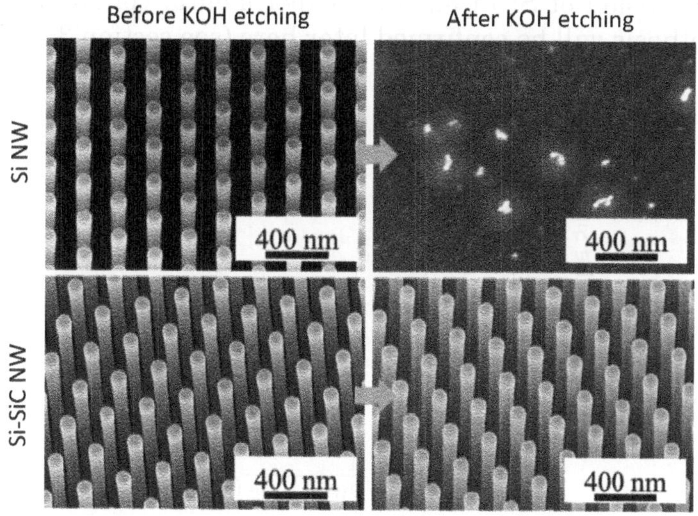

Figure 3.19. *SEM micrographs of Si and Si–SiC NW before and after the potash etching*

3.4.2.4. *Influence of the carburization time – kinetic study*

For an optimal carburization temperature of 1100°C, the dwell time is modified from 1 to 60 min. The thickness of the 3C–SiC layer is measured using STEM images at 30 keV or TEM images at 200 keV.

It would appear that the thickness of the cubic SiC shell formed, ε_{SiC}, varies linearly with the square root of the carburization time at 1100°C t_2 (see Figure 3.20). Carburization is therefore limited by a mechanism of diffusion. According to the previous studies of the carburization of Si substrates [CIM 00], we have seen that it is silicon, rather than carbon, that out-diffuses through SiC (see section 3.2). The observation of voids at the Si/SiC interfaces [FER 97] experimentally confirms this phenomenon in the literature. In our case, during this study, no voids at the Si/SiC interfaces were observed. We will assume that carburization is also governed by the out-diffusion of Si in our own experimental conditions; this hypothesis will be confirmed later here (see section 3.5).

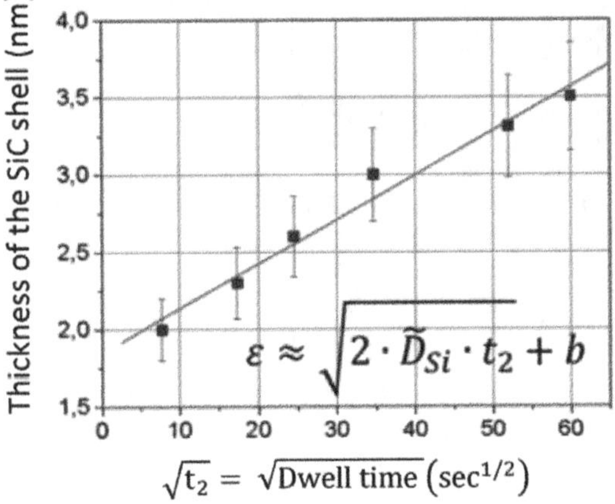

Figure 3.20. *Thickness of the SiC shell in function of the squared root of the carburization dwell time at 1100°C*

We can determine an apparent diffusion coefficient from the following Einstein–Smoluchowski equation on diffusion:

$$\varepsilon_{SiC} = \sqrt{2.D_{Si}.t_2 + b} \quad [3.1]$$

The apparent (in the experimental sense) diffusion coefficient determined, D_{Si}, is equal to 3.7×10^{-18} cm^2.s^{-1}, in accordance with the previous works on thin layers. Figure 3.21 – taken from Figure 3.6 – which draws on works by Cimalla *et al.* [CIM 00] (see section 3.2), groups the different values of the out-diffusion coefficients of Si in function of the growth techniques. The diffusion coefficient established in this study does indeed correspond to the chemical vapor deposition (CVD) growth of a 3C–SiC single crystalline layer in accordance with Arrhenius' law. It is interesting to remark that even at this nanometric scale, the diffusion remains Fickian. It must still be noted that the mechanisms of diffusion, at the atomic scale, of Si through SiC are not well-known, even to this day.

When carburization time t_2 at 1100°C is equal to zero, we can extrapolate the line and determine, as a first approximation, the thickness of the SiC formed between 800°C and 1100°C. We find that $b \approx 1.9$ nm. We hypothesize here that the seeding step is already governed by the out-diffusion of Si, which is a strong hypothesis, but gives an interesting order of magnitude.

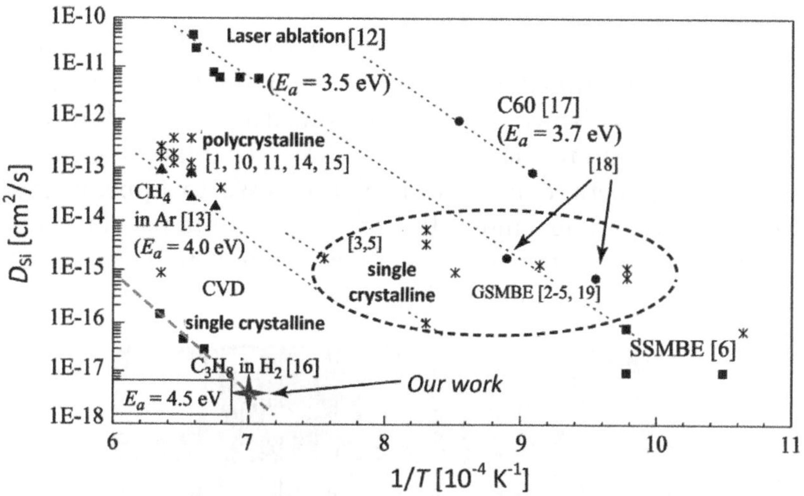

Figure 3.21. *Values of the out-diffusion of Si through SiC during the carburization step in function of temperature for different experiments, from Cimalla et al. [CIM 00]. The value of the out-diffusion coefficient determined in our experimental conditions is also marked*

3.4.2.5. *Carburization of Si NW in a high-purity reactor*

The carburization of Si NW has also been studied in a high-purity reactor dedicated to the hetero-epitaxy of 3C–SiC on Si. Without going into detail on this expitaxial apparatus, which is largely described in the following references

[FER 96a, FER 97], it is important to mention the notable differences with the previous reactor:

– heating is achieved by electromagnetic induction, with a maximal temperature ramp up of 10°C.s^{-1};

– the working pressure is still atmospheric pressure, with the possibility of injecting the gases C_3H_8, H_2, SiH_4 and Ar. The carburization phase is carried out with a dilution ratio of 1:1,000 of the propane in the dihydrogen, whose volume flow is equal to 16 L.min^{-1}.

Under these new experimental conditions, it is first of all most interesting to note that following the temperature rise (which is very quick in this case, and under a strong H_2 flow), the Si NWs are (deoxidized) smooth and facetted (over all the length up to the peak of the NW). This faceting causes an irregular octagonal section of the NWs to appear, seen from above. The mobility of the ad-atoms at high temperatures allows for this arrangement to occur while minimizing surface energy (Figure 3.22).

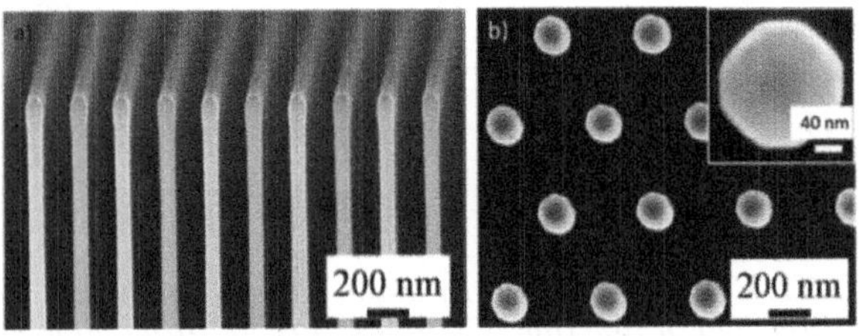

Figure 3.22. *SEM images of the long, smooth Si NWs at high temperature. a) Overall view at 90°. b) View from above, and in the box there is a close-up of the irregular octagonal section*

Following carburization, the morphology of the NWs seems unchanged *a priori*, but careful structural characterization (TEM observation of a transverse thin lamella of a NW) reveals that a cubic, single crystalline SiC shell with a thickness of 2 nm coats the Si NWs (see Figure 3.23). The Fourier transforms of the high-resolution TEM images show that the Si NWs are facetted along planes {220} and {400}. The {400} planes are more spread out than the {220} planes, which results in the irregular octagonal section observed.

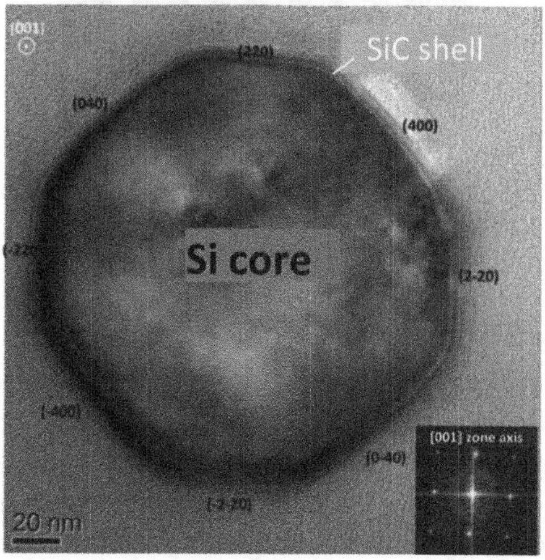

Figure 3.23. *TEM image of the transverse section of a smooth and carburized NW observed along [001] zone axis*

These experiments have demonstrated the growth of core–shell single-crystalline (for the core and the shell) Si–SiC NWs that are long, smooth and facetted along planes {220} and {400}.

The epitaxial regrowth, possible due to the availability of propane and silane gases installed in the reactor, is applied on these new objects: new core–(thick) shell NWs have thus been obtained (Figure 3.24). A section made through these NWs thanks to the ion beam in the FIB-SEM allows us to observe the Si core (octagonal section, dark gray) covered by a SiC shell (square section, light gray), see Figure 3.25. The anisotropy of the growth rates of SiC is clearly demonstrated here: from the measurements of the thickness of the SiC shell, a factor of 6 between the growth rates depending on the direction, {220} or {400} is measured.

Figure 3.24. *SEM image of Si nanowires after carburization and return of epitaxy*

3.4.2.6. *Conclusions and perspectives of the state of the art*

It would appear that the study of carburization of Si NWs leads to the growth of single-crystalline Si–SiC NWs [OLL 13], contrarily to previous works (see section 2.1.4, Figure 2.6). The combining of top-down approaches, for the realization of a system of single-crystalline Si NWs whose doping is easily controlled, with a bottom-up approach of carburization has led to the synthesis of a material of excellent morphological quality (controlled object size, smooth sidewall surfaces, dense and consistent shell), with a near-ideal level of crystalline quality.

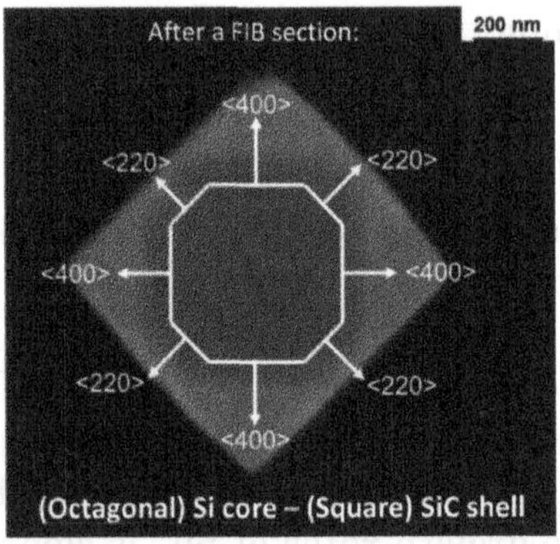

Figure 3.25. *SEM image of an epitaxial and carburized Si nanowire after sectioning using FIB. The nanowire is made up of a Si core with an irregular octagonal shape and a SiC shell whose external shape is a square section*

3.5. Growth of silicon carbide nanotubes

3.5.1. *Founding idea and experimental application*

According to the previous studies on the synthesis of core–shell Si–SiC NWs, the carburization of Si NWs also seems to be a promising process for the growth of SiC nanotubes. The out-diffusion of Si would need to be exacerbated until the Si core was completely empty and lead to the transformation of the Si NW into SiC NT. We wish here to use the Kirkendall effect, discovered as early as 1947 during the study of the diffusion of Cu and Zn in brass [FAN 07, KIR 42, SMI 47]. This effect is based on the difference between the diffusion of two elements (like Si and C here) through a solid (here SiC).

However, the time and temperature parameters are not sufficient here to significantly increase the out-diffusion of silicon[3] (see section 3.4.2.4). Using the Le Chatelier principle, the starting idea was to control out-diffusion via carburization pressure.

Thus, a three-step carburization technique was established:

– the Si NWs are first heated around atmospheric pressure, and the methane is injected, from 800°C to 1100°C at a flow rate of 0.5 mL.min^{-1} to avoid any sublimation or damage to the nanostructures;

– next, at the optimal carburization temperature, 1100°C, the methane flow is increased to 4 mL.min^{-1} and the carburization pressure is decreased (to 350 Pa);

– finally, the nanostructures are cooled to ambient temperature at close to atmospheric pressure.

3 At 1100°C, after 4 h of carburization carried out at atmospheric pressure, the thickness of the shell would be of 5 nm; and at 1200°C, under the same experimental conditions, it would only reach 7 nm.

Figure 3.26 summarizes the experimental conditions put in place to obtain SiC NTs.

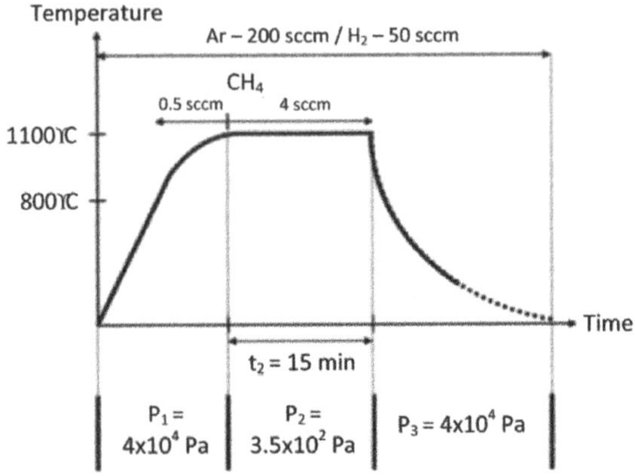

Figure 3.26. *Schematic representation of the thermal profile and the gas flows used in the SiC nanotubes growth*

Following this particular carburization, the morphology is this time remarkably modified, with the appearance of faceting of the nanostructures along a square section. A FIB section inside the nanostructures shows their tubular structure. Nanotubes – closed at their end – have indeed been synthetized (Figure 3.28).

Figure 3.27. *SEM image of short Si NW, carburized under reduced pressure: a) view from above, b) after FIB milling*

To identify the chemical and crystalline nature of the nanotubes, Raman spectroscopy is used successfully (unlike the previous studies which involved to small SiC thicknesses[4]), and demonstrates unambiguously the growth of 3C–SiC nanotubes. The full width at half maximum of the TO and LO mode peaks are relatively small (30 cm^{-1} for the TO mode) and seem to indicate a good crystalline quality of the SiC. To compare, the full width at half maximum of the TO mode is 4 cm^{-1} on a bulk single crystal of 3C–SiC (for which stacking faults density is in the order of 10^4 cm^{-1}) [BOU 06]. Stacking faults (SF) are undoubtedly present in the SiC nanotubes (see shouldering of the TO mode on the Raman spectrum). The observations made by transmission electron microscopy of a longitudinal thin lamella of these nanostructures have shown that these 3C–SiC nanotubes have dense sidewalls that are facetted along the {200} planes (Figures 3.28 and 3.29). The four planes that define the square section of the nanotubes are of the {200} type, just like the top of the nanotube. The interior base of the nanotube (toward the Si substrate) is itself facetted along the {220} type planes[5]. The thickness of the walls is between 40 and 100 nm depending on the height of the nanotube; the high sidewall of the nanotube is systematically the highest thickness according to the observations ($\approx$100 nm).

All of the characterizations presented here involve the carburization of initially short Si NWs, and this result can be transposed to long NWs; also transformed into 3C–SiC nanotubes, their structure remains tubular from the base to the top (see Figure 3.30).

---

4 The penetration depth of the laser in the visible spectrum is in the order of 2 mm in the SiC and 1 μm in the Si.

5 Let us note that if the starting Si nanowires were oriented along the ⟨111⟩ crystalline axis, rather than the ⟨100⟩ axis like in our study, it is highly likely that the orientation of the crystalline planes of the facets of the nanotubes would be different.

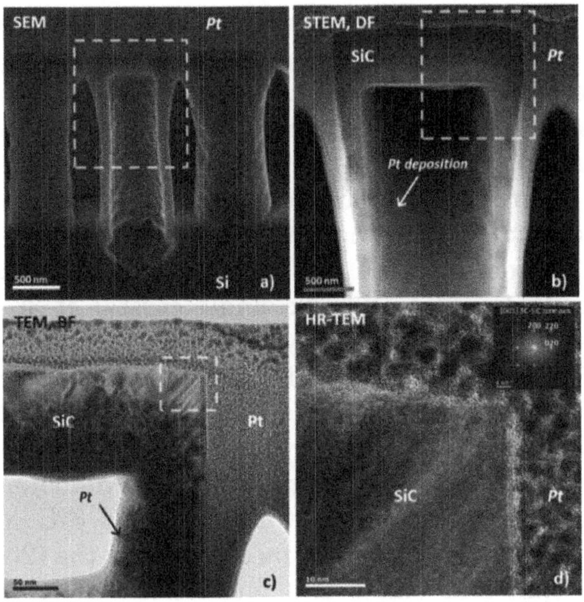

Figure 3.28. *Observation of a longitudinal thin lamella of a SiC nanotube. a) SEM image (secondary electrons). b) STEM image, dark field at 200 keV. c) STEM image, bright field at 200 keV. d) High resolution TEM image of the top corner of the nanotube observed along the [001] zone axis. Platinum is used as the protective deposit in the creation of the thin lamella in FIB-SEM, we can observe a re-deposit inside the nanotube*

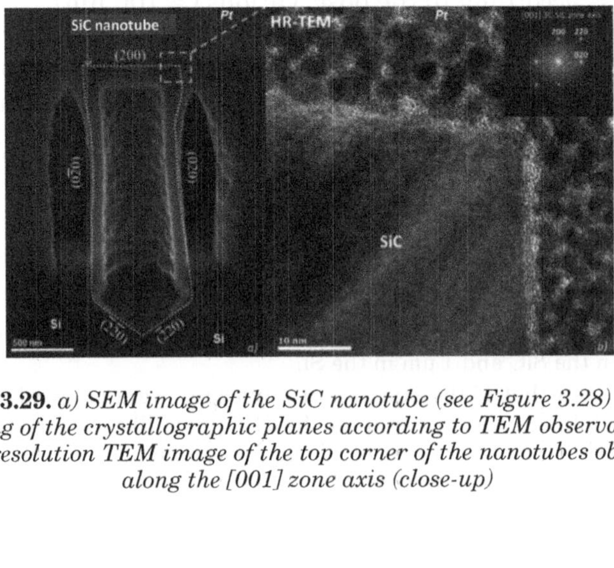

Figure 3.29. *a) SEM image of the SiC nanotube (see Figure 3.28) with the indexing of the crystallographic planes according to TEM observation. b) High resolution TEM image of the top corner of the nanotubes observed along the [001] zone axis (close-up)*

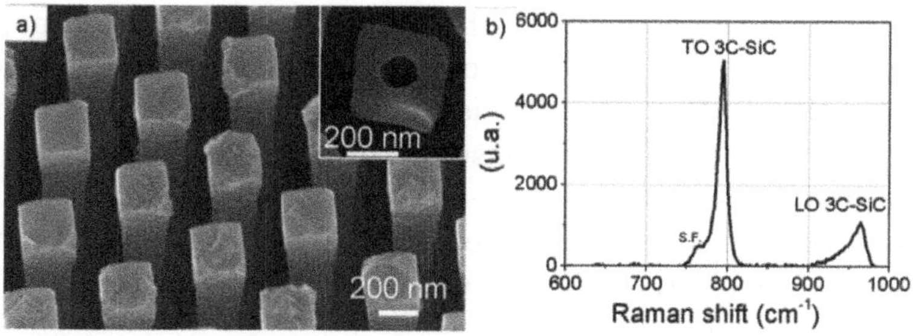

Figure 3.30. *a) SEM image of long SiC nanotubes and in the inset a micrograph of the nanotube seen from above following sectioning done in the FIB. b) Raman spectrum of the 3C–SiC nanotubes (laser wavelength: 514.5 nm; power of the laser measured on the sample: 0.1 mW)*

The sidewalls of the SiC nanotubes are dense, the crystalline quality is very good (as verified by the high-resolution TEM image from the top of a nanotube), and the dimensions of each nanotube are identical and controllable (following the initial dimensions of the Si NWs) [LAT 13].

While experimental conditions were modified from the previous study, the limiting step in the carburization remains the diffusion. These new experiments also confirm that it is indeed Si that out-diffuses through SiC. Pressure is an experimental parameter that allows us to control this out-diffusion very sensitively by influencing the concentration gradient (the diffusion flow being proportional to the concentrations gradient). This is why we propose to build a diagram showing the existence domains of different nanostructures obtained in the carburization of Si NWs (see section 3.6.2).

3.5.2. *A word on the kinetics of carburization*

For these new experimental conditions leading to the growth of nanotubes, the carburization time is decreased from 2 to 5 min: in this case, the Si NWs are transformed into core–shell Si–SiC NWs for which the thickness of the

SiC shell is respectively 12 nm and 27 nm (see STEM image, Figure 3.31). The out-diffusion coefficient of Si extrapolated from these two points is evaluated here as being 2×10^{-14} $cm^2.s^{-1}$. The increase of the diffusion coefficient in relation to the previous study is mainly due to a decrease in pressure (increase of the chemical potential gradient), and, to a lesser extent, to the presence of structural defects in the thick SiC shell.

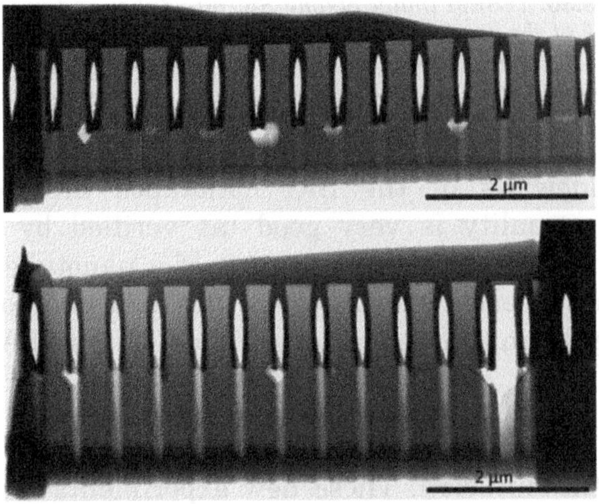

Figure 3.31. *STEM images at 30 keV of transverse thin lamella of carburized Si NW for: 2 min (above) and 5 min (below) at 1100°C and under carburization pressures of ($P_1 = 4 \times 10^4$ Pa and $P_2 = 3.5 \times 10^2$ Pa). Observation of voids at the bottom corner of the nanowires*

It is also very interesting to observe the thin lamellas corresponding to this type of sample as it allows us to study the initial steps of formation and to put forward ideas for the propagation pathway of the voids (Figures 3.31 and 3.32). According to the experimental conditions, the starting point of the void is either located at the foot of the NW at the level of the curve (Figure 3.31) or at the top corner (Figure 3.32) corner of the NW. In both of these scenarios, faults will probably be encountered that initiate the formation and propagation of voids at the level of these curves. At the foot of the NW, it could be a morphological fault caused by etching (micro-trenches), whereas at the top of the NW, a planar structural defects (stacking faults, for example) in the SiC shell at the intersection of the {200} type planes is more likely.

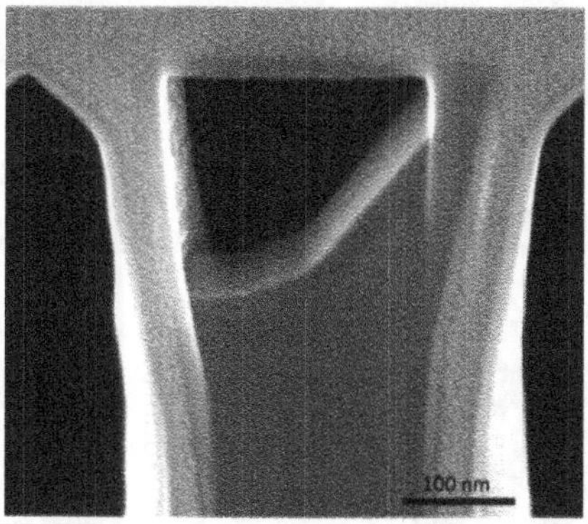

Figure 3.32. *SEM image at 5 kV of a transverse thin lamella of NW of Si carburized for 15 min at 1150°C and under carburization pressures equal to $P_1 = 9 \times 10^4$ Pa and $P_2 = 3.5 \times 10^2$ Pa. Observation of the void at the top corner of the nanowire*

3.6. Summary of the study of the carburization of silicon nanowires

3.6.1. *Illustration of carburization mechanisms for the growth of Si–SiC nanowires or SiC nanotubes*

The illustration below (Figure 3.33) illustrates the carburization mechanisms, which, depending on the experimental conditions, lead to either core–shell Si–SiC NWs or SiC nanotubes. First of all, the Si NWs are brought to high temperatures under dihydrogen at atmospheric pressure so as to reduce the layer of native oxide and prevent the sublimation of Si. From 800°C to 1100°C, the carbon precursor gas, methane, is injected into the carburization chamber and causes the formation of a thin shell of SiC (with a thickness in the order of 2 nm) which protects all of the nanostructures.

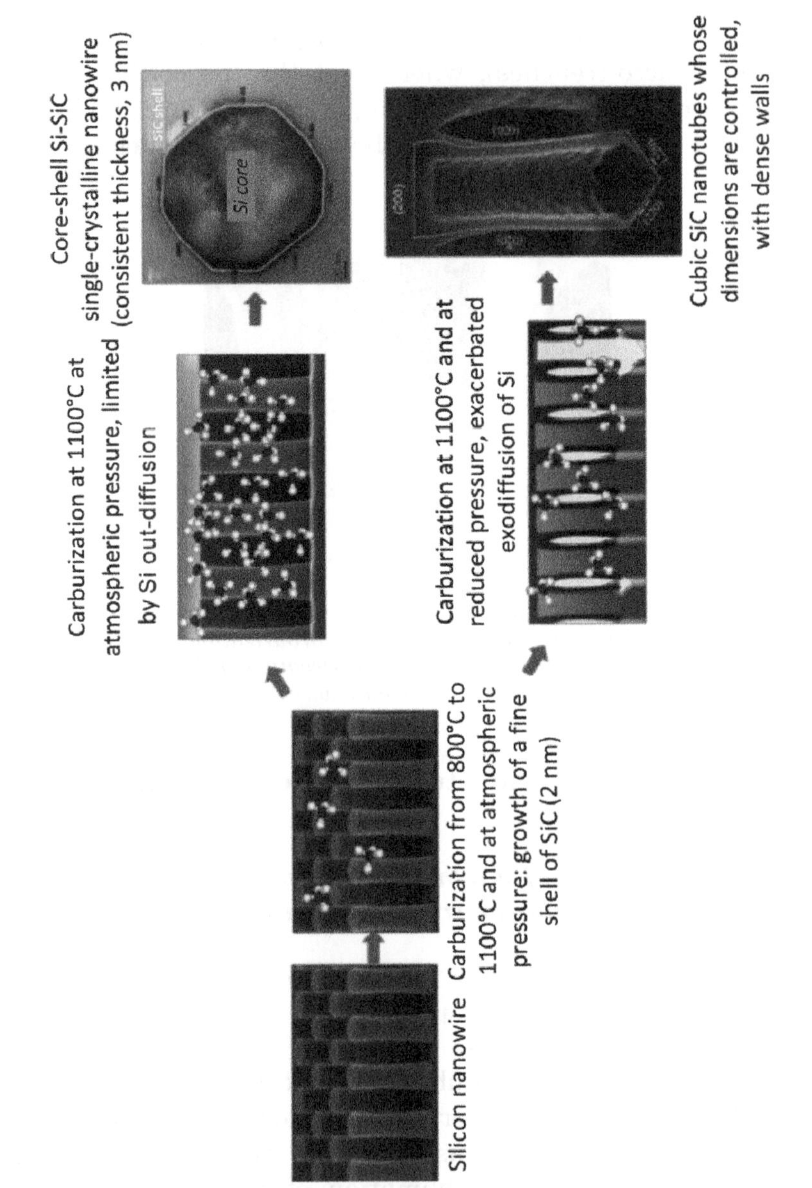

Figure 3.33. *Illustrations of the growth mechanisms of SiC during carburization depending on the experimental conditions, leading to the growth of Si–SiC nanowires or SiC nanotubes*

Next, depending on the type of nanostructures desired, the pressure at the carburization dwell – the length of which is around 15 min – at the optimal functioning temperature (here 1100°C) is modified. If the pressure at the carburization dwell is fixed at atmospheric pressure, core–shell Si–SiC single-crystalline NWs are grown. If the pressure at the carburization dwell is decreased to a few hundred Pa, SiC nanotubes are grown. In any case, it is the out-diffusion of Si through SiC that controls carburization. The speed of out-diffusion of Si is influenced notably by the pressure, which modifies the concentration gradient of chemical species.

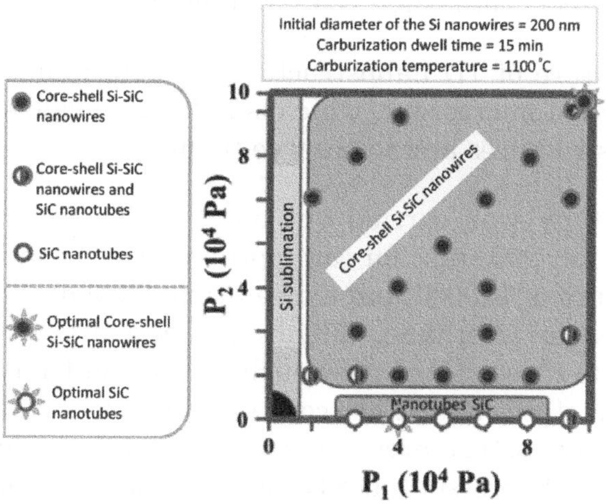

Figure 3.34. *Diagram of the existence of different forms of nanostructures obtained following the carburization of Si nanowires with an initial diameter of 200 nm, carried to 1100°C for 15 min following the pressures during the temperature rise (P_1) and during the carburization dwell (P_2)*

3.6.2. *The carburization of Si NW summarized: construction of an existence domain diagram*

We have just seen that the pressure during the temperature increase (written P_1) and during the carburation dwell (written P_2) were key parameters in the growth of one-dimensional SiC nanostructures. To give a more general sense to this study, a diagram showing the existence domains of different nanostructures is drawn experimentally (Figure 3.34).

On this diagram, the x-axis represents the value of the pressure P_1 whereas the value of the pressure at the carburization threshold P_2 is on the y-axis. In the (P_1, P_2) plane, the type of nanostructure obtained is noted:

– core–shell Si–SiC NWs;

– SiC nanotubes;

– or a mix of core–shell NWs and nanotubes.

The nature of 1D nanostructures is determined by using the FIB-SEM microscope, which is able to provide sections in the nanostructures and to observe them.

This study was carried out on Si NWs with an initial diameter of 200 nm for an optimal carburization dwell temperature of 1100°C and a dwell time fixed at 15 min. The flow rates for methane, dihydrogen and argon gases are identical to those stated previously (4 sccm of CH_4 diluted in 500 sccm of H_2 and 200 sccm of Ar). The availability of etched Si nanostructures – with identical and reproducible sizes – is still an essential advantage in carrying out such an experimental study.

Three regions of interest are apparent in this diagram:

– the region where the pressure during the temperature increase is low, causing damage to the nanostructures via the sublimation of Si;

– the regions where P_1 and P_2 are relatively high, leading to the growth of core–shell Si–SiC NWs;

– and finally the region where P_1 is high while P_2 is low, leading to the transformation of the Si NW into SiC NT.

Let us note that on this diagram the position of the optimal experimental points corresponding to the presentation of the previous results of the Si–SiC NWs and the SiC nanotubes is also indicated.

The adjustment of the (P_1 and P_2) couple allows the SiC shell to thicken considerably, between around 2 and 60 nm (see Figure 3.35), which again illustrates the importance of these parameters on the mechanisms that govern the out-diffusion of Si. *In fine*, the adjustment of the pressure allows us to greatly modify the speed of out-diffusion of Si and allows us to control the thickness of the SiC formed in a large range (from 2 nm for the shell to 100 nm for the sidewalls of the nanotubes); this speed of out-diffusion is far less sensitive to temperature changes (see Arrhenius' Law) or to the carburization time (see Figure 3.20).

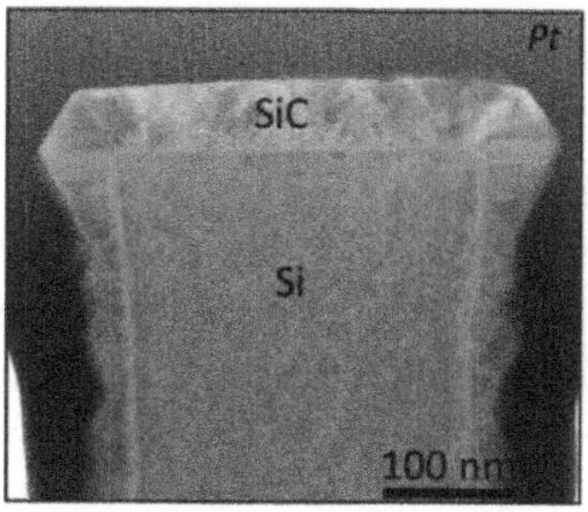

Figure 3.35. *STEM image at 30 kV of a transverse thin lamella of a core–shell Si–SiC nanowire with a very thick shell. The Si nanowire was carburized for 5 min at 1100°C for carburization pressures of* $P_1 = 6.7 \times 10^4$ *Pa and* $P_2 = 3.5 \times 10^2$ *Pa*

This diagram presents general experimental conditions for the synthesis of nanotubes and core–shell NWs, and shows that there are no extensive regions that favor the synthesis of an entirely carburized SiC NW. In other words, the growth of a SiC NW of a hundred nanometers in diameter would appear to be very difficult using this process.

However, we have been able to obtain a SiC NW by adjusting the pressure during the carburization threshold and by decreasing the initial diameter of the Si NWs. In this case, the Si NWs are first synthesized via the bottom-up approach and the VLS mechanism (catalyzed by gold colloids of calibrated sizes). The SiC NW is entirely carburized, and as is often the case for these materials that possess a zinc-blende or diamond structure, the direction of growth is of the ⟨111⟩ type. However, many stacking faults, also distributed in the {111} dense planes have been observed by TEM (Figure 3.36).

3.6.3. *Criticism of the nanostructures obtained*

The carburization of Si NWs previously obtained by plasma etching has led to the growth of two types of original nanostructure.

The core–shell Si–SiC NWs are single-crystalline and the dimensions of the NWs are controlled (length, external diameter, thickness of the SiC shell). Moreover, the doping of the core of the Si is easily controlled via the adequate etching of Si substrates. Compared to the literature (see Chapter 2, Figure 2.8), these results are completely novel.

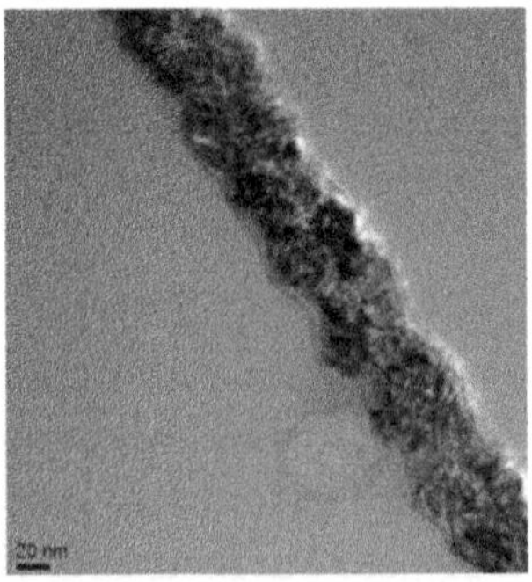

Figure 3.36. *TEM image (at 200 kV) of a SiC nanowire with a diameter of 50 nm (entirely carburized Si nanowire). The direction of growth is identified as the <111> type*

While the SiC nanotubes synthesized in the literature suffer from the presence of pores in their walls and from heterogeneity in terms of crystalline and morphological quality (see Chapter 2, Figure 2.7), the carburization of the NW of Si systems results in SiC NT of good crystalline quality, with dense walls, and once again reproducible and controllable dimensions[6].

These results are due to both the excellent starting bases that are etched Si NWs, and to the control of the carburization of Si NWs via the control of the out-diffusion of Si.

It is interesting to be able to compare this mixed top-down/bottom-up (etching/carburization) approach with a purely top-down approach of the etching of bulk SiC or in thin layers. Indeed, as with all other growth methods, the creation of stacking faults inside the SiC NWs could not be avoided with the carburization of Si NWs. To get around this problem, we shall study a second growth process through an entirely top-down approach.

6 SiC NT are closed at the top-end. A first experimental work showed that it is possible to "open" them using plasma etching (Ar/SF_6). Sulfur hexafluoride leads to the formation of radicals that attack the re-deposition of SiC induced by the physical etching of Ar plasma.

4

SiC-Based One-dimensional Nanostructure Technologies

The goal of this chapter is to consider the different usable technologies that involve silicon carbide (SiC). We shall find that it is possible to operate using a purely top-down approach (by plasma etching) to produce good-quality SiC nanowires (NWs). The possible applications for the different one-dimensional (1D) SiC nanostructures are then reviewed. Using our knowledge and the available literature, four main fields of application can be highlighted [LAT 14].

4.1. Top-down approach: SiC plasma etching for the production of SiC nanowires

The electronic and crystalline quality of SiC in its bulk form – obtained by sublimation or in thin layers, usually by chemical vapor deposition (CVD) epitaxy – allow us to consider the top-down approach as an adequate method for the production of 1D SiC nanostructures.

The plasma etching of SiC is known and here we have called upon already established knowledge. The etching of NWs with high aspect ratio, however, has not yet been mastered. Studies on anisotropic etching are essentially carried out to produce micro-trenches with small aspect ratios for micro-electro-mechanical systems (MEMS) [LEE 99, CHA 01]. To etch the SiC, which is a hard material, inductive plasma etching with SF_6/O_2 chemistry would appear to be more adequate [PLA 03]. The fundamental principle behind plasma-assisted SiC NWs etching is very similar to the one behind the etching of Si NWs, which was presented in section 3.1. However, it must be noted that the experimental parameters, as well as the nature of the etching mask and the chemical species in the plasma vary.

The chemical reactions that allow the gas SF_6 to etch SiC will first of all consist of the generation of reactive species in the plasma. Next, these reactive species break the Si–C bonds using different mechanisms, thus forming SiF_x and CF_y species that are evacuated by pumping [YIH 97].

After studying the different experimental parameters (dilution, etching time, material used for the mask, polarization of the substrate, etc.), it appears that SiC NWs can be made [CHO 12a] with a maximal aspect ratio of 7. Due to their morphology, these NWs shall be called nanopillars hereafter. The shape of these nanopillars is larger at the base compared to the peak, and obtaining vertical sidewalls would appear difficult. This “Eiffel Tower” morphology is explained by the re-deposition of etching products (of the SiF_xO_y type) during the etching process. The higher part of the pillar is continuously exposed to ionic bombardment; the nickel mask is eroded and the top becomes thinner, leading to non-vertical profiles. To deal with this problem, the Bosch process, which associates etching steps with passivation steps, could be used next.

As the etching time increases, the shape of the etched SiC nanopillar changes. For short etching times, its shape is round as seen from above, similar to the shape of the circular mask. As the etching time increases, the SiC nanopillar becomes facetted (Figure 4.1). The appearance of crystalline planes helps minimize the free surface energy (Wulff's theorem).

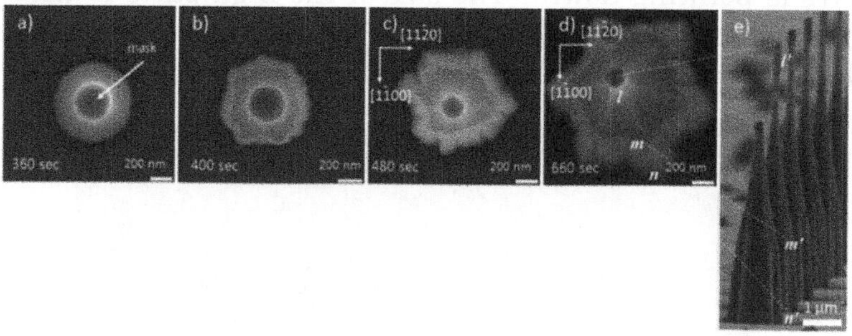

Figure 4.1. *SEM image MEB (a–d) seen from top view and e) side view of the 4H–SiC (0001) nanopillars after etching time equals to: a) 360, b) 400, c) 480 and (d–e) 660 sec*

The etching of several polytypes (4H, 6H, 15R and 3C) with different crystalline orientations has been studied [CHO 12b]. This study shows the presence of facetted nanopillars in hexagonal, rhombohedral or triangular sections (see Figure 4.2). For example, for the <111>-oriented 3C-SiC, we can observe on the sidewalls of the nanopillars the other equivalent {111} planes that are the densest planes that are the densest planes and those with the lowest free surface energy.

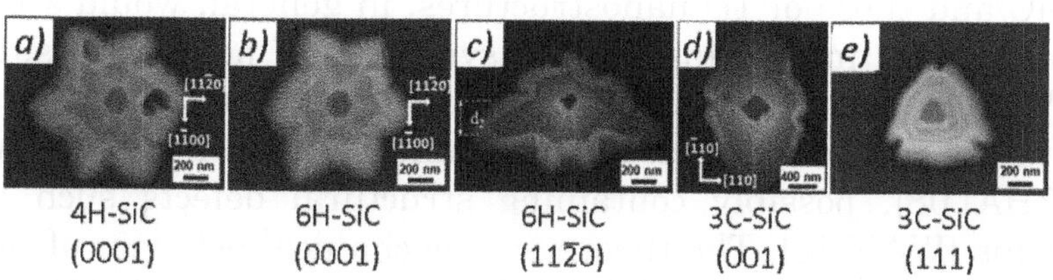

Figure 4.2. *SEM image of the top view of nanopillars of different SiC polytypes obtained after a long etching time: a) 4H–SiC (0001), b) 6H–SiC (0001), c) 6H–SiC $(11\bar{2}0)$, d) 3C–SiC (001) and e) 3C–SiC (111)*

In our study, the top of the SiC nanopillar has a diameter of approximately 100 nm in the last micrometer (Figure 4.3).

The higher part of these nanopillars is recovered via ultrasounds, with the substrate placed so that the "head of the NWs is pointing down". The size of the high parts is enough to result in the production of NW transistors to study their electronic properties.

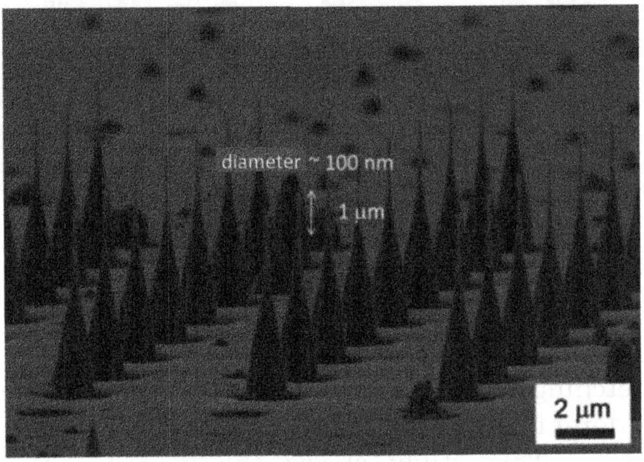

Figure 4.3. *SEM image of SiC nanopillars obtained using SF_6/O_2 plasma etching*

4.2. Mechanics

The combination of the excellent mechanical properties of SiC and those of 1D nanostructures, in general, would seem potentially interesting for the synthesis of new composites [VIV 06]. Based on this idea, in the literature, many *ab-initio* studies can be found on the mechanical behavior of SiC NWs [ZHA 08], possibly containing structural defects such as twins [WAN 10]. The theoretical mechanical behavior of SiC nanotubes in tension [LAT 14] and in torsion [JIA 11] has also been studied. Experimental studies are less common, but a significant experimental result is the observation of superplastic behavior in a SiC NW (with a diameter of $\approx$100 nm) with a deformation measured up to 200% in uniaxial traction (see Figure 4.4).

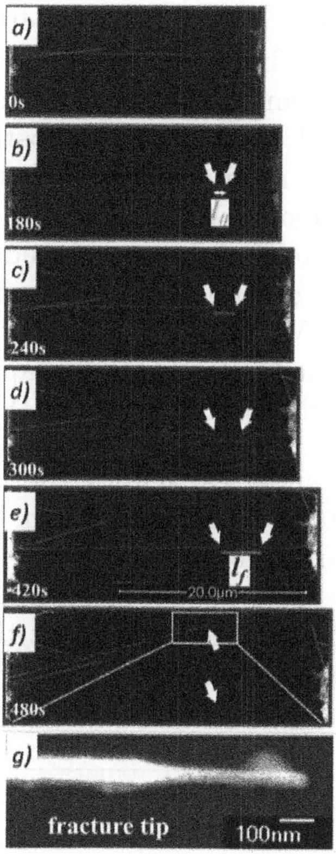

Figure 4.4. *a)–f) SEM images taken in-situ during an elongation test for a SiC nanowire. g) High-resolution image by back-scattered electrons of the tip of a SiC nanowire after fracture. From [ZHA 07]*

4.3. Energy

Due to the chemical inertness of SiC, the use of this material also seems adapted for energy storage in the construction of super-capacitors or Li-ion batteries.

Several advantages have shown to exist for using Si NW as the anode material in Li-ion batteries. Silicon has an intrinsically low discharge potential and has the theoretically highest known charge capacity (4,200 $mAh.g^{-1}$). However, after insertion and extraction of Li, the material undergoes an expansion in volume of 400%. The use of Si NW instead of the bulk material allows us to partly overcome this problem, and some excellent performances have been achieved with this nanostructured material [CHA 08]. However, after several cycles, the deformed Si NWs become amorphous. We think that the core–shell Si–SiC NW systems that we have developed could be useful in increasing the life of these batteries by conserving the integrity of the Si NW. Recently, the use of an anode made up of Si NWs decorated with SiC and graphene nanoparticles has actually provided promising results [YAN 13].

In the domain of super-capacitors, the work by Alper *et al.* [ALP 12] deserves to be mentioned again. They demonstrated the stability during more than 1,000 cycles of a super-capacitor made up of Si–NW covered with a polycrystalline layer of SiC (see Figure 2.8(d)); without the SiC deposit, the super-capacitor only operated once. The SiC deposit notably limits parasitic oxidation reactions (see Figure 4.5).

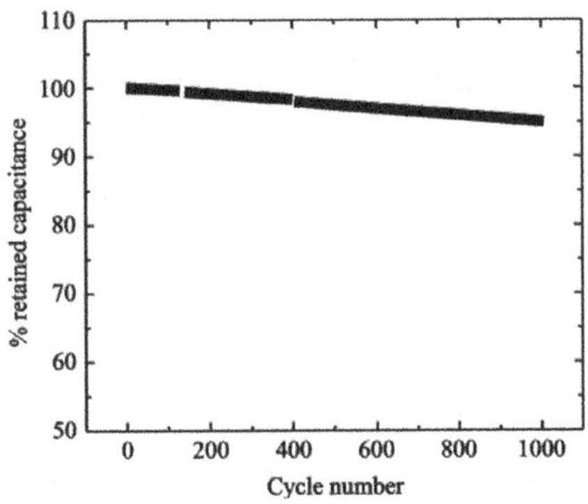

Figure 4.5. *Electric capacitance retained by the electrode made up of silicon nanowires covered in SiC. A total of 95% of the capacitance is retained after 1,000 cycles. From [ALP 12]*

As previously mentioned, the use of SiC nanotubes is also considered for the storage of dihydrogen; this is validated by theoretical considerations.

4.4. Electronics

4.4.1. *Integration of nanostructures in a nanowire transistor*

We have chosen to create NW-Field Effect Transistors (NWFETs) using "simple" technology. The three types of nanostructures that we produced (core–shell Si–SiC NWs, SiC nanotubes and SiC nanopillars) become the FET channel. This transistor is represented in Figure 4.6 as it was made.

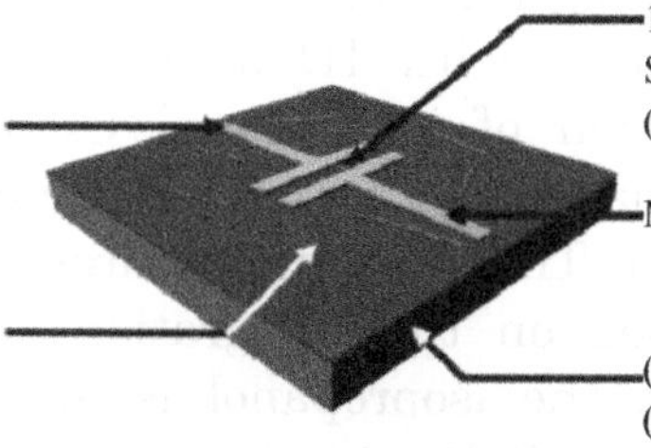

Figure 4.6. *Schematic representation of a nanowire field-effect nano-transistor*

Even if the fabrication of these NW-based transistors is one of the simplest forms of production, it involves several technological steps. This preliminary study demonstrates the quality of the material and its potential applications with a functional set-up [OLL 13].

Three kinds of transistor have thus been fabricated: i) the Si NW transistor, which is the reference for this study (the Si NWs are those used for carburization), ii) the core–shell Si–SiC NW transistor and iii) the SiC nanotube transistor.

Figure 4.7 goes back to the different technological steps that lead to the fabrication of a field-effect nano-transistor with a 1D planar nanostructure channel:

1) Starting with a (111)-oriented p^{++}-type Si substrate, available commercially, the first step consists of depositing a

layer of silicon nitride Si_3N_4 by CVD with a thickness of ≈200 nm (see Figure 4.7(a)). The Si_3N_4 layer can also be replaced by a layer of SiO_2 obtained by thermal oxidation of the substrate;

2) Once this step has been done, the substrate is coated with a protective resin, and cut into pieces of approximately 1 cm^2. The protective resin as well as the dusts generated during cutting are eliminated by chemical attack in acetone;

3) The next step consists of depositing the 1D nanostructures to be integrated onto the surface of the Si_3N_4. To do this, the substrate with 1D nanostructures is first submerged in a solution of isopropanol $CH_3CH(OH)–CH_3$ (IPA), after which the nanostructures are disengaged from the substrate by sonication. These nanostructures in solution in IPA are then dispersed on the integration substrate (see Figure 4.7(c)). Finally, the isopropanol is evaporated on a heating plate (see Figure 4.7(d));

4) The spin coating of a negative photoresist is the first step of the definition of the metallic contacts of the drain and of the source (see Figure 4.7(e));

5) This is followed by the alignment of the lithography mask above a nanostructure by optical control. The lithography mask defines the length of the channel. In our case, the distance between drain and source is 1 μm (see Figure 4.7(f));

6) Once the mask is aligned, the non-protected photoresist is insulated by a deep ultra-violet emitting light, 365 nm (see Figure 4.7(g));

7) The non-insulated photoresist is then removed using a specific solution, and the sample is submitted to a dioxygen plasma to clean the contacts. Finally, a solution of hydrofluoric acid is used to deoxidize the 1D nanostructure before the metal is deposited (see Figure 4.7(h));

8) A metallic layer made up of 120 nm of nickel and 100 nm of aluminum is then deposited by evaporation. This layer is

deposited in the patterns defined by lithography and thus forms the contacts (see Figure 4.7(i));

9) The sample is then submerged in a solution of acetone to dissolve the non-isolated photoresist, which causes the metal deposited above this resin to disappear. This step, called "lift-off", is usually followed by a final cleaning plasma so as to make the contacts cleaner (see Figure 4.7(j)). The nano-transistor is finally finished.

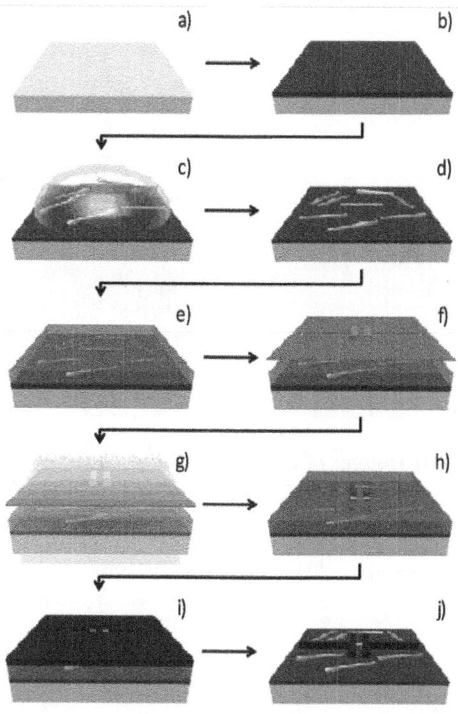

Figure 4.7. *The main steps of the production of a field-effect nano-transistor with a 1D nanostructure as channel. a) p_{++}-degenerated Si substrate. b) Deposition of a layer of Si_3N_4 serving as the gate oxide. c) Deposition of a drop of isopropanol containing 1D nanostructures. d) Evaporation of the isopropanol to deposit the nanostructures on the surface. e) Spin-coating of a negative photoresist. f) Alignment of a lithography mask defining the drain and the source. g) Insulation by photolithography. h) Removal of the insulated photoresist by a revelator. i) Deposition of the contacts by evaporation of Ni/Al; j) Removal of the metal deposited on the resin by "lift-off"*

In Figure 4.8, scanning electron microscope (SEM) images are shown of a sample of p_{++}-degenerated Si(100) on which a field-effect nano-transistor has been built.

The source–drain current/voltage characteristics for a gate voltage varying between –20 and +20 V of the Si and Si–SiC NW and SiC nanotube transistors are presented in Figure 4.9.

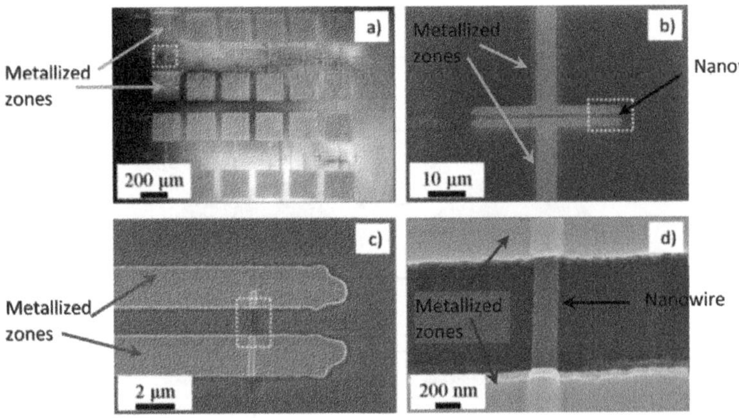

Figure 4.8. *SEM images of the Si nanowire channel field-effect nano-transistor observed at different levels of magnification*

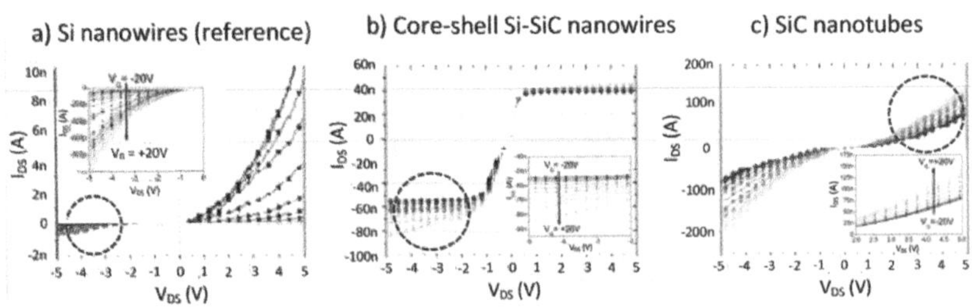

Figure 4.9. *Measuring of the current between the drain and the source for different voltages imposed between the drain and the source and for a gate voltage varying between –20 and +20 V of three back gate nanowire nano-transistors: a) Si nanowire, b) core–shell Si–SiC nanowire and c) SiC nanotube. In the box, there is a close-up of the measurement of current*

A transistor effect is observed with an effect of the gate voltage on the values of the current in the drain, which is used for the three structures. However, the apparent mobilities of the carriers (here the electrons) taken from the electric characteristics of the "Si NW", "core–shell Si–SiC NW" and "SiC nanotube" nano-transistors are very low, in the order of 0.01 cm^2.V^{-1}.s^{-1}. It must be noted that no annealing of the contacts is carried out initially. It is commonly observed that after silicidation of the contacts, the behavior of the transistor can be greatly improved. By extrapolating the gains brought about by the silicidation of the contacts to the Si NW transistors (via appropriate annealing) [ROS 11], the optimization of the electric contacts should result in reasonable performance levels. A rapid thermal annealing is also necessary to activate the electronic properties [MAT 04] of the nanostructures produced.

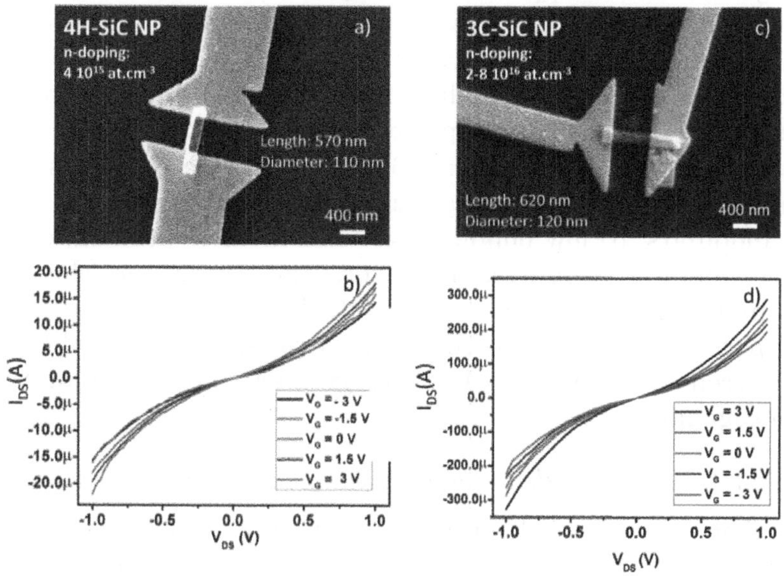

Figure 4.10. *a) and b) SEM images of nano-transistors made from nanopillars of a) 4H–SiC and c) 3C–SiC. b) and d) Measurement of the current between the drain and the source for different voltages imposed between the drain and the source and for a gate voltage varying between –3 and +3 V: b) for the 4H–SiC channel and d) for the 3C–SiC channel*

Nanowire transistors have also been made from SiC nanopillars obtained by etching. The difficulty of such an integration must be noted for SiC nanopillars whose length is limited to a few hundred nanometers (for example the need to proceed to an additional step of electronic–beam lithography). Two SiC polytypes, 3C-(001) and 4H-(0001), both with relatively low doping levels (10^{15}–10^{16} atoms.cm^{-3}) have been integrated in these transistors (Figure 4.10). The low doping level is chosen with the goal of improving the mobility of the carriers [ROS 01]. Compared to the previous study, an optimized annealing of the Ni contacts was successfully executed (at 650°C for 30 sec under N_2), with the goal of forming silicides (Ni_xSi_y) at the interface between the nickel and the SiC [HOO 13].

The current levels measured in these new transistors are higher, and the apparent mobility of the electrons is in the order of 40 and 140 cm^2.V^{-1}.s^{-1}, respectively, for the 4H–SiC and 3C–SiC transistors [HOO 13], which is a record in this domain [OLL 13]. The highest measurement of mobility made in SiC NW transistors reported in the literature before now was of 16 cm^2.V^{-1}.s^{-1} [ZHO 06]. Our excellent results are mainly due to i) the high crystalline quality of the SiC nanopillars, ii) low doping levels and iii) the optimization of the metallic contacts. However, the transistor channel cannot be completely closed yet. We can note that our experimental results seems to confirm that the mobility of electrons in the cubic form is greater than those in the hexagonal form.

4.4.1.1. *With regards to the method of extraction of mobilities*

The method used for the extraction of mobilities is the same as the methods used in the literature and corresponds to the determination of μ, the apparent mobility. The transconductance g of the transistor (whose channel has a length of L) is first of all determined by measuring the derivative of the current in the drain with respect to the polarization of the gate for a low value of voltage between the drain and the source V_{DS}:

$$g = \frac{\partial I_{DS}}{\partial V_G} \quad [4.1]$$

Transconductance is itself a function of mobility and of the oxide capacity of gate C_{ox} (estimated by a calculation taking into account the geometry of the transistor), according to the relation:

$$\mu = \frac{g.L^2}{V_{DS}C_{ox}} \quad [4.2]$$

The oxide capacity of the gate is calculated following the "cylinder on plane" model, according to this equation:

$$C_{ox} = \frac{2\pi\varepsilon_0\varepsilon_{ox}L}{cosh^{-1}\left(\frac{r+e_{ox}}{r}\right)} \quad [4.3]$$

where ε_0 represents the permittivity of the vacuum, ε_{ox} the permittivity of the oxide, r the radius of the 1D nanostructure and e_{ox} the thickness of the gate oxide.

4.5. For biology

Based on the electronic, chemical and biocompatibility properties of SiC, many applications are envisaged in the domains of biotechnology and biomedicine [YAK 07]. SiC could be used in cardiovascular or neurological systems, or even in cell staining [SAD 11]. A first step was made in studying the grafting of deoxyribonucleic acid (DNA) onto SiC nanopillars and Si–SiC NWs that we produced. The grafting of DNA observed by fluorescence (see Figure 4.11) shows that DNA is grafted preferably on the sidewalls of the nanopillars, confirming the advantage of obtaining large surfaces developed with these nanostructures [FRA 14].

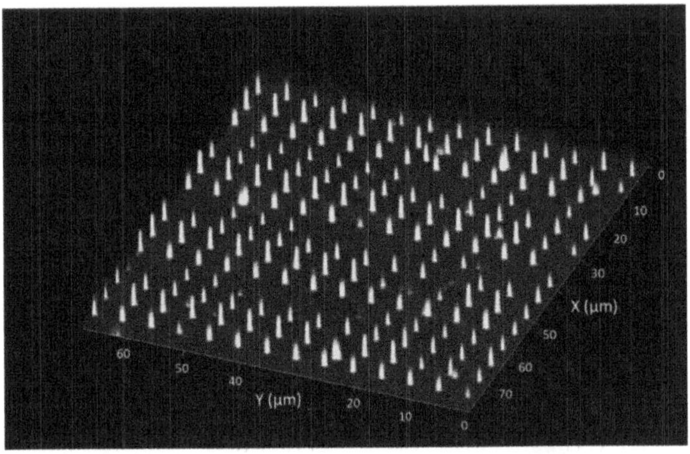

Figure 4.11. *3D reconstruction by confocal microscopy of the fluorescence of DNA on the SiC nanopillars. From [FRA 14]*

4.6. Future work

SiC 1D nanostructure-based nano-transistors and nano-sensors could become innovative nano-devices. Currently, only SiC NW-based transistors have been made. Electronic performances remain low, far from the theoretical performances. This is due, in part, to the low crystalline quality of the material, overly high non-controlled doping and finally the difficulty of integrating the NW (the production of contacts). We were able to measure an apparent mobility in the order of $140\ cm^2.V^{-1}.s^{-1}$, or approximately 10 times more than the previous record [LEE 99]. However, we are still far from the ideal operation of a transistor. This research topic is still relatively new and progress will undoubtedly be made in the next few years.

The use of SiC nanotubes is currently limited to *ab-initio*-type theoretical studies that demonstrate their potential use as chemical detectors. A definite advantage of SiC nanotubes that we have grown (whose relatively size exempts them from possible quantum or organizational surface effects) is that they benefit from similar electronic properties, unlike C nanotubes, which are both conductors and semiconductors.

Conclusion

The topic of this work was the presentation of one-dimensional (1D) silicon carbide-based nanostructures. While they remain relatively unstudied in the literature compared to 1D silicon-based nanostructures, SiC-based nanostructures are obviously attractive in their ability to combine in a single object the remarkable properties of silicon carbide and the unique properties of nano-dimensional objects.

Thus, it is high levels of thermal conductivity, biocompatibility or even a high breakdown field that the scientific community seeks to ally with the high surface to volume ratio of nanowires, and of other nanotubes. However, the experimental reality presents more of a contrast, as seen in the results obtained from the different growth methods used for SiC nanostructures in the literature. This is why we have made the choice of presenting one promising process: the carburization of silicon nanowires.

This encouraging pathway combines nanotechnology's symbolic top-down and bottom-up approaches, so as to benefit from both nearly ideal silicon nanowires and the growth of SiC by addition of carbon atoms in the gaseous phase.

In this way our research has focused on studying the material: how to grow it, and its physical and chemical properties, alternatively and iteratively, so as to obtain the highest crystalline quality possible. This study has also led to the production of the first electronic devices that use our nanostructures.

The carburization of Si nanowires proved to be particularly interesting. While carburization is globally well-known in Si substrates, its use in nanowires constitutes original work. Good understanding of the physical and chemical mechanisms at work during carburization has led to relatively new results, show-casing the growth of core–shell Si–SiC nanowires and SiC nanotubes of very high crystalline quality (Figure C.1).

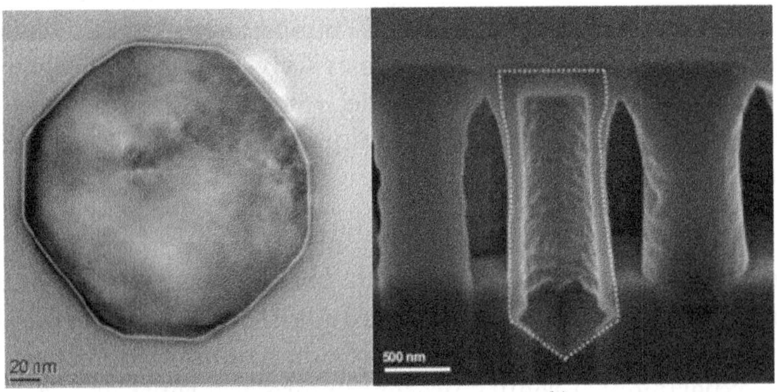

a) b)

Figure C.1. *a) TEM image of a Si–SiC nanowire disc and b) SEM image of a SiC NT*

Many promising future applications have come out of this work. The complete study of the doping of these nanostructures (intrinsic, n type, p type) still needs to be carried out. Notably, the effect of carburization on the controlled doping of the core in Si still needs to be evaluated.

An important work of optimization in nanostructure integration technologies is also necessary to study the electronic properties of the material. In particular, a lot of work must be done in the optimization of contacts by silicidation. Furthermore, the electronic properties of SiC should be activated by high-temperature post-production annealing.

Bio-functionalization and its effects on components are of sizable importance, aiming to demonstrate the added value of silicon carbide.

The growth of suspended graphene from core–shell Si–SiC nanowire systems or SiC nanotubes is probably worth studying.

Other applications based on the use of these Si–SiC NW or SiC NT networks certainly deserved to be looked into (electrodes operating in harsh environments, etc.).

Bibliography

Chapter 1

[AGA 13] AGARWAL A., SINGH R., RYU S.H., *et al.*, 600 V, 1-40 A, Schottky diodes in SiC and their applications, Technical report, Cree Inc., 2013.

[ALG 08] ALGRA R.E., VERHEIJEN M.A., BORGSTROM M.T., *et al.*, "Twinning superlattices in indium phosphide nanowires", *Nature*, vol. 456, no. 7220, pp. 369–372, 2008.

[BAI 07] BAI X., GUO X., MCINTOSH D.C., *et al.*, "High detection sensitivity of ultraviolet 4H–SiC avalanche photodiodes", *IEEE Journal of Quantum Electronics*, vol. 43, no. 12, pp. 1159–1162, 2007.

[BAU 15] BAUMHAUER H., "Über die Kristalle des Carborundums", *Zeitschrift fur Kristallographie und Mineralogie*, vol. 55, pp. 33–39, 1915.

[BEC 97] BECHSTEDT F., KACKELL P., ZYWIETZ A., *et al.*, "Polytypism and properties of silicon carbide", *Physica Status Solidi B – Basic Research*, vol. 202, no. 1, pp. 35–62, 1997.

[BRO 93] BROWN D.M., DOWNEY E.T., GHEZZO M., *et al.*, "Silicon carbide UV photodiodes", *IEEE Transactions on Electron Devices*, vol. 40, no. 2, pp. 32–333, 1993.

[CAM 98] CAMASSEL J., CONTRERAS S., ROBERT J.S., "Matériaux semi-conducteurs à grand gap: SiC", *Les Techniques de l'Ingénieur*, vol. E1990, pp. 1–12, 1998.

[CAM 08] CAMBAZ Z.G., YUSHIN G., OSSWALD S., *et al.*, "Noncatalytic synthesis of carbon nanotubes, graphene and graphite on SiC", *Carbon*, vol. 46, pp. 841–849, 2008.

[CHO 04] CHOI H.J., SEONG H.K., LEE J.C., *et al.*, "Growth and modulation of silicon carbide nanowires", *Journal of Crystal Growth*, vol. 268, pp. 472–478, 2004.

[END 97] ENDO M., IIJIMA S., DRESSELHAUS M.S., *Carbon Nanotubes*, Pergamon Press, 1997.

[FEN 10] FENG X., LATEMPA T.J., BASHAM J.I., *et al.*, "Ta_3N_5 nanotube arrays for visible light water photoelectrolysis", *Nano Letters*, vol. 10, no. 3, pp. 948–952, 2010.

[FER 06] FERRO G., Croissance épitaxiale du carbure de silicium à basse température, HDR, Claude Bernard University – Lyon 1, 2006.

[GOL 06] GOLDBERGER J., HOCHBAUM A.I., FAN R. *et al.*, "Silicon vertically integrated nanowire field effect transistors" *Nano Letters*, vol. 6, no. 5, pp. 973–977, 2006.

[HOC 05] HOCHBAUM A.I., FAN R., HE R., *et al.*, "Controlled growth of Si nanowire arrays for device integration", *Nano Letters*, vol. 5, no. 3, pp. 457–460, 2005.

[HUA 01] HUANG M.H., WU Y., FEICK H., *et al.*, "Catalytic growth of zinc oxide nanowires by vapor transport", *Advanced Materials*, vol. 13, no. 2, pp. 113–116, 2001.

[HU 08] HU J., XIN X., LI X., *et al.*, "4H–SiC visible-blind single-photon avalanche diode for ultraviolet detection at 280 and 350 nm", *IEEE Transactions on Electron Devices*, vol. 55, no. 8, pp. 1977–1983, 2008.

[HU 09] HU M., YU R., MACMANUS-DRISCOLL J.L., *et al.*, "Large-area silica nanotubes with controllable geometry on silicon substrates", *Applied Surface Science*, vol. 255, no. 6, pp. 3563–3566, 2009.

[IIJ 91] IIJIMA S., "Helical microtubules of graphitic carbon", *Nature*, vol. 354, pp. 56–58, 1991.

[IOF 13] Ioffe Physico-Technical Institute, New semiconductor materials, characteristics and properties, 2013, available at: www.ioffe.ru/SVA/NSM/introduction.html.

[JOH 02] JOHNSON J.C., CHOI H.J., KNUTSEN K.P., *et al.*, "Single gallium nitride nanowire lasers", *Nature Materials*, vol. 1, no. 2, pp. 106–110, 2002.

[KAC 07] KACIULIS S., PANDOLFI L., COMINI E., *et al.*, "Nanowires of metal oxides for gas sensing applications", *Surface and Interface Analysis*, vol. 40, nos. 3–4, pp. 575–578, 2007.

[KAP 85] KAPLAN R., WAGNER R.J., KIM H.J., *et al.*, "Electron cyclotron resonance in cubic SiC", *Solid State Communications*, vol. 55, no. 1, pp. 67–69, 1985.

[KUM 03] KUMAR K.S., VAN SWYGENHOVEN H., SURESH S., "Mechanical behavior of nanocrystalline metals and alloys", *Acta Materialia*, vol. 51, no. 19, pp. 5743–5774, 2003.

[LAI 96] LAI S.L., GUO J.Y., PETROVA V., *et al.*, "Size-dependent melting properties of small tin particles: nanocalorimetric measurements", *Physical Review Letters*, vol. 77, pp. 99–102, 1996.

[LAT 06] LATU-ROMAIN L., Croissance de monocristaux massifs de carbure de siliciumcubique, PhD Thesis, Institut National Polytechnique de Grenoble, 2006.

[LAU 02] LAUHON L.J., GUDIKSEN M.S., WANG C.L., *et al.*, "Epitaxial core-shell and core-multishell nanowire heterostructures", *Nature*, vol. 420, no. 6911, pp. 57–61, 2002.

[LI 06] LI Y., QIAN F., XIANG J., *et al.*, "Nanowire electronic and optoelectronic devices", *Materials Today*, vol. 9, no. 10, pp. 18–27, 2006.

[MER 09] MERCIER F., Cristallogenèse de carbure de silicium cubique en solution à haute température, PhD Thesis, Institut National Polytechnique de Grenoble, 2009.

[MIL 06] MILLER J.T., KROPF A.J., ZHA Y., *et al.*, "The effect of gold particle size on Au-Au bond length and reactivity toward oxygen in supported catalysts", *Journal of Catalysis*, vol. 240, no. 2, pp. 222–234, 2006.

[MOH 96] MOHAMMAD S.N., MORKOÇ H., "Progress and prospects of group–III nitride semiconductors", *Progress in Quantum Electronics*, vol. 20, nos. 5–6, pp. 361–525, 1996.

[MOM 11] MOMMA K., IZUMI F., "VESTA for three-dimensional visualization of crystal, volumetric and morphology data", *Journal of Applied Crystallography*, vol. 44, no. 6, pp. 1272–1276, 2011.

[MOR 94] MORKOÇ H., STRITE S., GAO G.B., *et al.*, "Large-band-gap SiC, III-V nitride, and II-VI ZnSe-based semiconductor-device technologies", *Journal of Applied Physics*, vol. 76, no. 3, pp. 1363–1398, 1994.

[MOR 07] MOR G.K., PRAKASAM H.E., VARGHESE O.K., *et al.*, "Vertically oriented Ti-Fe-O nanotube array films: toward a useful material architecture for solar spectrum water photoelectrolysis", *Nano Letters*, vol. 7, no. 8, pp. 2356–2364, 2007.

[NAL 02] NALLET F., "SiC pour l'électronique de puissance du futur", *Les Techniques de l'Ingénieur*, vol. RE3, pp. 1–11, 2002.

[NIL 99] NILSSON H.E., HJELM M., "Monte Carlo simulation of electron transport in 2H–SiC using a three valley analytical conduction band model", *Journal of Applied Physics*, vol. 86, no. 11, pp. 6230–6233, 1999.

[NIQ 00] NIQUET Y.M., DELERUE C., ALLAN G., *et al.*, "Method for tight-binding parametrization: application to silicon nanostructures", *Physical Review B*, vol. 62, no. 8, pp. 5109–5116, 2000.

[OST 05] OSTOJIC G.N., ZARIC S., KONO J., *et al.*, "Stability of high-density one-dimensional excitons in carbon nanotubes under high laser excitation", *Physical Review Letters*, vol. 94, p. 097401, 2005.

[PAR 09] PARK M.H., KIM M.G., JOO J., *et al.*, "Silicon nanotube battery anodes", *Nano Letters*, vol. 9, no. 11, pp. 3844–3847, 2009.

[PER 04] PERSSON A.I., LARSSON M.W., STENSTROM S., *et al.*, "Solid-phase diffusion mechanism for GaAs nanowire growth", *Nature Materials*, vol. 3, no. 10, pp. 677–681, 2004.

[POT 11] POTIÉ A., BARON T., LATU-ROMAIN L., *et al.*, "Controlled growth of SiGe nanowires by addition of HCl in the gas phase", *Journal of Applied Physics*, vol. 110, no. 2, p. 024311, 2011.

[RAM 47] RAMSDELL L.S., "Studies on silicon carbide", *American Mineralogist*, vol. 32, nos. 1–2, pp. 64–82, 1947.

[RAY 07] RAYNAUD C., "Propriétés physiques et électroniques du carbure de silicium (SiC)", *Les Techniques de l'Ingénieur*, vol. D3119, pp. 1–14, 2007.

[SAD 11a] SADDOW S.E., *Silicon Carbide Biotechnology: A Biocompatible Semiconductor for Advanced Biomedical Devices and Applications*, Elsevier Science, 2011.

[SAD 11b] SADDOW S.E., FREWIN C.L., COLETTI C., *et al.*, "Single-crystal silicon carbide: a biocompatible and hemocompatible semiconductor for advanced biomedical applications", in MONAKHOV E.V., HORNOS T., SVENSSON B.G. (eds.), *Silicon Carbide and Related Materials 2010*, vol. 679–680 of *Materials Science Forum*, pp. 824–830, 2011.

[SAU 04] SAUER R., TEOFILOV N., THONKE K., "Exciton condensation in diamond", *Diamond & Related Materials*, vol. 13, nos. 4–8, pp. 691–699, 2004.

[SCH 55] SCHNEER C.J., "Polymorphism in one dimension", *Acta Crystallographica*, vol. 8, no. 5, pp. 279–285, 1955.

[SCH 05] SCHMIDT V., SENZ S., GÖSELE U., "Diameter-dependent growth direction of epitaxial silicon nanowires", *Nano Letters*, vol. 5, no. 5, pp. 931–935, 2005.

[SCH 09] SCHMIDT V., WITTEMANN J.V., SENZ S., *et al.*, "Silicon nanowires: a review on aspects of their growth and their electrical properties", *Advanced Materials*, vol. 21, pp. 2681–2702, 2009.

[SEL 01] SELLMYER D.J., ZHENG M., SKOMSKI R., "Magnetism of Fe, Co and Ni nanowires in self-assembled arrays", *Journal of Physics-Condensed Matter*, vol. 13, no. 25, pp. R433–R460, 2001.

[SHA 69] SHAFFER P.T.B., "A review of the structure of silicon carbide", *Acta Crystallographica Section B*, vol. 25, no. 3, pp. 477–488, 1969.

[SON 94] SON N.T., KORDINA O., KONSTANTINOV A.O., *et al.*, "Electron effective masses and mobilities in high-purity 6H–SiC chemical vapor deposition layers", *Applied Physics Letters*, vol. 65, no. 25, pp. 3209–3211, 1994.

[SON 95] SON N.T., CHEN W.M., KORDINA O., *et al.*, "Electron effective masses in 4H–SiC", *Applied Physics Letters*, vol. 66, no. 9, pp. 1074–1076, 1995.

[THI 44] THIBAULT N.W., "Morphological and structural crystallography and optical properties of silicon carbide (SiC)", *The American Mineralogist*, vol. 29, pp. 249–278, 1944.

[TSA 05] TSAKALAKOS L., RAHMANE M., LARSEN M., *et al.*, "Mo_2C nanowires and nanoribbons on Si by two-step vapor-phase growth", *Journal of Applied Physics*, vol. 98, no. 4, p. 044317, 2005.

[WAG 64] WAGNER R.S., ELLIS W.C., "Vapor-liquid-solid mechanism of single crystal growth", *Applied Physics Letters*, vol. 4, no. 5, pp. 89–90, 1964.

[WEN 03] WEN J.G., LAO J.Y., WANG D.Z., *et al.*, "Self-assembly of semiconducting oxide nanowires, nanorods, and nanoribbons", *Chemical Physics Letters*, vol. 372, pp. 717–722, 2003.

[WU 01] WU Y., YANG P., "Direct observation of vapor-liquid-solid nanowire growth", *Journal of the American Chemical Society*, vol. 123, pp. 3165–3166, 2001.

[WU 05] WU X.L., FAN J.Y., QIU T., *et al.*, "Experimental evidence for the quantum confinement effect in 3C–SiCN anocrystallites", *Physical Review Letters*, vol. 94, p. 026102, 2005.

[YOF 01] YOFFE A.D., "Semiconductor quantum dots and related systems: electronic, optical, luminescence and related properties of low dimensional systems", *Advances in Physics*, vol. 50, no. 1, pp. 1–208, 2001.

[YOS 95] YOSHIDA S., *Properties of Silicon Carbide*, HARRIS G.L. (ed.), 1995.

[ZHA 02] ZHANG H.F., WANG C.M., WANG L.S., "Helical crystalline SiC/SiO_2 core-shell nanowires", *Nano Letters*, vol. 2, no. 2, pp. 941–944, 2002.

[ZHA 99] ZHANG Y., ICHIHASHI T., LANDREE E., *et al.*, "Heterostrucutres of single-walled carbon nanotubes and carbide nanorods", *Science*, vol. 285, pp. 1719–1722, 1999.

Chapter 2

[ALP 12] ALPER J.P., VINCENT M., CARRARO C., *et al.*, "Silicon carbide coated silicon nanowires as robust electrode material for aqueous micro-supercapacitor", *Applied Physics Letters*, vol. 100, no. 16, p. 163901, 2012.

[ATT 08] ATTOLINI G., ROSSI F., BOSI M., *et al.*, "Synthesis and characterization of 3C–SiC nanowires", *Journal of Non-Crystalline Solids*, vol. 354, pp. 5227–5229, 2008.

[ATT 09] ATTOLINI G., ROSSI F., FABBRI F., *et al.*, "A new growth method for the synthesis of 3C–SiC nanowires", *Materials Letters*, vol. 63, pp. 2581–2583, 2009.

[BAN 11] BANERJEE S., NIGAM S., PILLAI C.G.S., *et al.*, "Hydrogen storage on Ti decorated SiC nanostructures: a first principles study", *International Journal of Hydrogen Energy*, vol. 37, no. 4, pp. 3733–3740, 2012. *International Conference on Renewable Energy (ICRE)*, Jaipur, India, January 17–21, 2011.

[BEA 11] BEABER A.R., GIRSHICK S.L., GERBERICH W.W., "Dislocation plasticity and phase transformations in Si–C core–shell nanotowers", *International Journal of Fracture*, vol. 171, pp. 177–183, 2011.

[BEC 05] BECHELANY M., CORNU D., CHASSAGNEUX F., *et al.*, "Chemical surface transformation of SiC-based nanocables", *Journal of Optoelectronics and Advanced Materials*, vol. 8, no. 2, pp. 638–642, 2006.

[BEC 07a] BECHELANY M., BRIOUDE A., CORNU D., *et al.*, "A Raman spectroscopy study of individual SiC nanowires", *Advanced Functional Materials*, vol. 17, pp. 939–943, 2007.

[BEC 07b] BECHELANY M., BRIOUDE A., STADELMANN P., *et al.*, "Very long SiC-Based coaxial nanocables with tunable chemical composition", *Advanced Functional Materials*, vol. 17, no. 16, pp. 3251–3257, 2007.

[CHA 06] CHAUSSENDE D., BLANQUET E., BAILLET F., *et al.*, "Thermodynamic aspects of the growth of SiC single crystals using the CF-PVT process", *Chemical Vapor Deposition*, vol. 12, pp. 541–548, 2006.

[CHE 12] CHEN Y., WANG H.X, ZHAO J.X., *et al.*, "Theoretical insights into the effects of the diameter and helicity on the adsorption of formic acid on silicon carbide nanotube", *Journal of Nanoparticle Research*, vol. 14, no. 1, pp. 1–11, 2012.

[CHO 04] CHOI H.J., SEONG H.K., LEE J.C., *et al.*, "Growth and modulation of silicon carbide nanowires", *Journal of Crystal Growth*, vol. 268, pp. 472–478, 2004.

[DAI 95] DAI H., WONG E.W., LU Y.Z., *et al.*, "Synthesis and characterization of carbide nanorods", *Nature*, vol. 375, pp. 769–772, 1995.

[FAB 10] FABBRI F., ROSSI F., ATTOLINI G., *et al.*, "Enhancement of the core near-band-edge emission induced by an amorphous shell in coaxial one-dimensional nanostructure: the case of SiC/SiO_2 core/shell self-organized nanowires", *Nanotechnology*, vol. 21, no. 345702, pp. 1–7, 2010.

[FAN 12] FANG J., AHARONOVICH I., LEVCHENKO I., *et al.*, "Plasma-enabled growth of single-crystalline SiC/AlSiC core-shell nanowires on porous alumina templates", *Crystal Growth and Design*, vol. 12, no. 6, pp. 2917–2922, 2012.

[FEN 12] FENG W., MA J., YANG W., "Precise control on the growth of SiC nanowires", *CrystEngComm*, vol. 14, no. 4, pp. 1210–1212, 2012.

[GAN 11] GANJI M.D., SEYED-AGHAEI N., TAGHAVI M.M., *et al.*, "Ammonia adsorption on SiC nanotubes: a density functional theory investigation", *Fullerenes, Nanotubes, and Carbon Nanostructures*, vol. 19, no. 4, pp. 289–299, 2011.

[HON 00] HONG M.H., SAMANT A.V., PIROUZ P., "Stacking fault energy of 6H–SiC and 4H–SiC single crystals", *Phylosophical Magazine a-Physics of Condensed Matter Structure Defects and Mechanical Properties*, vol. 80, no. 4, pp. 919–935, 2000.

[HU 04] HU J.Q., BANDO Y., ZHAN J.H., *et al.*, "Fabrication of ZnS/SiC nanocables, SiC-shelled ZnS nanoribbons (and sheets), and SiC nanotubes (and tubes)", *Applied Physics Letters*, vol. 85, no. 14, pp. 2932–2934, 2004.

[HYU 09] HYUN J.S., NAM S.H., KANG B.C., *et al.*, "Growth of 3C–SiC nanowires on nickel coated Si(100) substrate using dichloromethylvinylsilane and diethylmethylsilane by MOCVD method", *Physica Status Solidi C*, vol. 6, no. 4, pp. 810–812, 2009.

[KEL 03] KELLER N., PHAM-HUU C., EHRET G., *et al.*, "Synthesis and characterisation of medium surface area silicon carbide nanotubes", *Carbon*, vol. 41, pp. 2131–2139, 2003.

[KAN 03] KANG B.C., LEE S.B., BOO J.H., "Growth of β–SiC nanowires on Si(100) substrates by MOCVD using nickel as a catalyst", *Thin Solid Films*, vol. 464, pp. 215–219, 2004.

[KIM 03] KIM H.Y., BAE S.Y., KIM N.S., *et al.*, "Fabrication of SiC-C coaxial nanocables: thickness control of C outer layers", *Chemical Communications*, no. 20, pp. 2634–2635, 2003.

[KNI 68] KNIPPENBERG W.F., VERSPUI G., "Crystal growth of silicon carbide (III)", *Philips Technical Review*, vol. 29, nos. 8–9, 1968.

[KRI 11] KRISHNAN B., THIRUMALAI R.V.K.G., KOSHKA Y., *et al.*, "Substrate-dependent orientation and polytype control in SiC nanowires grown on 4H–SiC substrates", *Crystal Growth & Design*, vol. 11, no. 2, pp. 538–541, 2011.

[LAT 14] LATU-ROMAIN L., OLLIVIER M., "Silicon carbide-based one-dimensional nanostructures growth: towards electronics and biology perspectives", *Journal of Physics D: Applied Physics*, vol. 47, p. 203001, 2014. Topicalreview.

[LI 03] LI Z.J., LI H.J., CHEN X.L., *et al.*, "Large-scale synthesis of crystalline β-SiC nanowires", *Applied Physics A*, vol. 76, pp. 637–640, 2003.

[LI 05] LI Y., DOROZHKIN P.S., BANDO Y., *et al.*, "Controllable modification of SiC nanowires encapsulated in BN nanotubes", *Advanced Materials*, vol. 17, no. 5, pp. 545–549, 2005.

[LI 07] LI Y.L., ISHIGAKI T., "Synthesis and structural characterization of core-shell Si–SiC composite particles by thermal plasma in-flight carburization of silicon powder", *Journal of the Ceramic Society of Japan*, vol. 115, no. 1347, pp. 717–723, 2007.

[LIN 03] LINDEFELT U., IWATA H., OBERG S., *et al.*, "Stacking faults in 3C–, 4H–, and 6H–SiC polytypes investigated by an ab initio supercell method", *Physical Review B*, vol. 67, no. 15, p. 155204, 2003.

[MEN 07] MENG A., LI Z., ZHANG J., *et al.*, "Synthesis and Raman scattering of β–SiC/SiO_2 core-shell nanowires", *Journal of Crystal Growth*, vol. 308, no. 2, pp. 263–268, 2007.

[MO 02] MO Y.H., SHAJAHAN M., LEE K.S., *et al.*, "Growth and characterization of SiC micro-crystals on Si(100) substrate by the chemical vapor deposition method", *Diamond & Related Materials*, vol. 11, pp. 1703–1708, 2002.

[MPO 06] MPOURMPAKIS G., FROUDAKIS G.E., LITHOXOOS G.P., *et al.*, "SiC nanotubes: a novel material for hydrogen storage", *Nano Letters*, vol. 6, no. 8, pp. 1581–1583, 2006.

[NAZ 14] NAZARUDIN N.F.F.B., Azizan S.N.A.B., Rahman S.A., *et al.*, "Growth and structural property studies on NiSi/SiC core-shell nanowires by hot-wire chemical vapor deposition", *Thin Solid Films*, 2014.

[NHU 02] NHUT J.M., VIEIRA R., PESANT L., *et al.*, "Synthesis and catalytic uses of carbon and silicon carbide nanostructures", *Catalysis Today*, vol. 76, no. 1, pp. 11–32, 2002.

[PAN 00] PAN Z., LAI H.L., FREDERICK C.K., *et al.*, "Oriented silicon carbide nanowires: synthesis and field emission properties", *Advanced Materials*, vol. 12, no. 16, pp. 1186–1190, August 2000.

[PEI 06] PEI L.Z., TANG Y.H., CHEN Y.W., *et al.*, "Preparation of silicon carbide nanotubes by hydrothermal method", *Journal of Applied Physics*, vol. 99, no. 11, p. 114306, 2006.

[PHA 01] PHAM-HUU C., KELLER N., EHRET G., *et al.*, "The first preparation of silicon carbide nanotubes by shape memory synthesis and their catalytic potential", *Journal of Catalysis*, vol. 200, no. 2, pp. 400–410, 2001.

[SOU 13] SOUKIASSIAN P., WIMMER E., CELASCO E., *et al.*, "Hydrogen-induced nanotunnel opening within semiconductor subsurface", *Nature Communications*, vol. 4, November 2013.

[SUN 02] SUN X.H., LI C.P., WONG W.K., *et al.*, "Formation of silicon carbide nanotubes and nanowires via reaction of silicon (from disproportionation of silicon monoxide) with carbon nanotubes", *Journal of the American Chemical Society*, vol. 124, pp.14464–14471, 2002.

[SUN 07] SUNDARESAN S.G., DAVYDOV A.V., VAUDIN M.D., *et al.*, "Growth of silicon carbide nanowires by a microwave heating-assisted physical vapor transport process using group VIII metal catalysts", *Chemical Materials*, vol. 19, pp. 5531–5537, 2007.

[TAG 05] TAGUCHI T., IGAWA N., YAMAMOTO H., *et al.*, "Preparation and characterization of single-phase SiC nanotubes and C-SiC coaxial nanotubes", *Physica E: Low-Dimensional Systems and Nanostructures*, vol. 28, no. 4, pp. 431–438, 2005.

[TSA 05] TSAKALAKOS L., RAHMANE M., LARSEN M., *et al.*, "Mo_2C nanowires and nanoribbons on Si by two-step vapor-phase growth", *Journal of Applied Physics*, vol. 98, no. 4, p. 044317, 2005.

[TSA 06] TSAKALAKOS L., FRONHEISER J., ROWLAND L., *et al.*, "SiC nanowires by silicon carburization", *Materials Research Society Symposium Proceedings*, Cambridge University Press, pp. 0963–Q11-03, 2006.

[WAG 64] WAGNER R.S., ELLIS W.C., "Vapor-liquid-solid mechanism of single crystal growth", *Applied Physics Letters*, vol. 4, no. 5, pp. 89–90, 1964.

[WAL 08] WALLIS K.L., PATYK J.K., ZERDA T.W., "Reaction kinetics of nanostructured silicon carbide", *Journal of Physics: Condensed Matter*, vol. 20, no. 32, p. 325216, 2008.

[WAN 08] WANG H., XIE Z., YANG W., *et al.*, "Morphology control in the vapor-liquid-solid growth of SiC nanowires", *Crystal Growth & Design*, vol. 8, no. 11, pp. 3893–3896, 2008.

[WAN 10] WANG H., LIN L., YANG W., *et al.*, "Preferred orientation of SiC nanowires induced by substrates", *The Journal of Chemical Physics C*, vol. 114, pp. 2591–2594, 2010.

[WEI 08] WEI G., QIN W., WANG G., *et al.*, "The synthesis and ultraviolet photoluminescence of 6h–sic nanowires by microwave method", *Journal of Physics D: Applied Physics*, vol. 41, no. 23, p. 235102, 2008.

[WU 11] WU P., DUAN W.D., ZHAO J.X., *et al.*, "Theoretical study on the possibility of SiC nanotube as gas Sensors for HCHO detection", *Journal of Computational and Theoretical Nanoscience*, vol. 8, no. 12, pp. 2469–2473, 2011.

[XI 06] XI G., LIU Y., LIU X., *et al.*, "Mg-catalyzed autoclave synthesis of aligned silicon carbide nanostructures", *The Journal of Physical Chemistry B*, vol. 110, pp. 14172–14178, 2006.

[YAN 05a] YANG W., ARAKI H., TANG C., *et al.*, "Single-crystal SiC nanowires with a thin carbon coating for stronger and tougher ceramic composites", *Advanced Materials*, vol. 17, pp. 1519–1523, 2005.

[YAN 05b] YANG W., ARAKI H., THAVEETHAVORN S., *et al.*, "In situ synthesis and characterization of pure SiC nanowires on silicon wafer", *Applied Surface Science*, vol. 241, pp. 236–240, 2005.

[YAN 08] YANG Y., MENG G., LIU X., *et al.*, "Aligned SiC porous nanowire arrays with excellent field emission properties converted from Si nanowires on silicon wafer", *The Journal of Physical Chemistry C*, vol. 112, no. 51, pp. 20126–20130, 2008.

[ZHA 00] ZHANG Y.F., TANG Y.H., ZHANG Y., *et al.*, "Deposition of carbon nanotubes on Si nanowires by chemical vapor deposition", *Chemical Physics Letters*, vol. 330, pp. 48–52, 2000.

[ZHA 09a] ZHAO H., SHI L., LI Z., *et al.*, "Silicon carbide nanowires synthesized with phenolic resin and silicon powders", *Physica E*, vol. 41, pp. 753–756, 2009.

[ZHA 09b] ZHAO J.X., DING Y.H., "Can silicon carbide nanotubes sense carbon dioxide?", *Journal of Chemical Theory and Computation*, vol. 5, no. 4, pp. 1099–1105, 2009.

[ZHO 00] ZHOU X.T., ZHANG R.Q., PENG H.Y., *et al.*, "Highly efficient and stable photoluminescence from silicon nanowires coated with SiC", *Chemical Physics Letters*, vol. 332, nos. 3–4, pp. 215–218, 2000.

Chapter 3

[BAG 09] BAGCI S., DUMAN S., TUTUNCU H.M., *et al.*, "Ab initio calculations of phonon dispersion relations for bulk and surface of cubic", *Journal of Physics and Chemistry of Solids*, vol. 70, no. 2, pp. 444–450, 2009.

[BHA 98] BHATTI Q.A., MATTHAI C.C., "Critical thickness and growth modes of SiC layers on Si substrates – a molecular dynamics study", *Applied Surface Science*, vol. 123, pp. 7–10, 1998.

[BOU 06] BOULLE A., CHAUSSENDE D., LATU-ROMAIN L., *et al.*, "X-ray diffuse scattering from stacking faults in thick 3C-SiC single crystals", *Applied Physics Letters*, vol. 89, no. 9, p. 091902, 2006.

[CIM 00] CIMALLA V., WOHNER T., PEZOLD T.J., "The diffusion coefficient of silicon in thin SiC layers as a criterion for the quality of the grown layers", in CARTER C.H., DEVATY R.P., ROHRER G.S. (eds.), *Silicon Carbide and Related Materials – 1999*, vol. 338–3 of *Materials Science Forum*, pp. 321–324, 2000.

[DEN 08] DEN HERTOG M.I., ROUVIERE J.L., DHALLUIN F., *et al.*, "Control of gold surface diffusion on Si nanowires", *Nano Letters*, vol. 8, no. 5, pp. 1544–1550, 2008.

[FAN 07] FAN H.J., GÖSELE U., ZACHARIAS M., "Formation of nanotubes and hollow nanoparticles based on kirkendall and diffusion processes: a review", *Small*, vol. 3, no. 10, pp. 1660–1671, 2007.

[FER 06] FERRO G., Croissance épitaxiale du carbure de silicium à basse température, HDR, Claude Bernard University – Lyon 1, 2006.

[FER 96a] FERRO G., MONTEIL Y., VINCENT H., *et al.*, "Infrared kinetic study of ultrathin SiC buffer layers grown on Si(100) by reactive chemical vapour deposition", *Thin Solid Films*, vol. 278, pp. 22–27, 1996.

[FER 96b] FERRO G., MONTEIL Y., VINCENT H., *et al.*, "Atomic force microscopy growth modeling of SiC buffer layers on Si(100) and quality optimization", *Journal of Applied Physics*, vol. 80, pp. 4691–4702, 1996.

[FER 97] FERRO G., Mécanismes de croissance cristalline par épitaxie en phase vapeur du carbure de silicium cubique sur substrats de silicium orientés (100), PhD Thesis, Claude Bernard University – Lyon 1, 1997.

[GHO 66] GHOSHTAGORE R.N., COBLE R.L., "Self-diffusion in silicon carbide", *Physical Review*, vol. 143, pp. 623–626, 1966.

[GRA 72] GRAUL J., WAGNER E., "Growth mechanism of polycrystalline β–SiC layers on silicon substrate", *Applied Physics Letters*, vol. 21, no. 2, pp. 67–69, 1972.

[KIR 42] KIRKENDALL E.O., "Diffusion of zinc in alpha brass", *Transactions of the American Institute of Mining, Metallurgical and Petroleum Engineers*, vol. 147, pp. 104–109, 1942.

[LAT 08] LATU-ROMAIN L., MOUCHET C., CAYRON C., *et al.*, "Growth parameters and shape specific synthesis of silicon nanowires by the VLS method", *Journal of Nanoparticle Research*, vol. 10, pp. 1287–1291, 2008.

[LAT 11] LATU-ROMAIN L., OLLIVIER M., MANTOUX A., *et al.*, "From Si nanowire to SiC nanotube", *Journal of Nanoparticle Research*, vol. 13, pp. 5425–5433, 2011.

[LAT 13] LATU-ROMAIN L., OLLIVIER M., THINEY V., *et al.*, "Silicon carbide nanotubes growth: an original approach", *Journal of Physics D: Applied Physics*, vol. 46, no. 9, p. 092001, 2013.

[LI 95] LI J.P., STECKL A.J., "Nucleation and void formation mechanisms in SiC thin film growth on Si by carbonization", *Journal of the Electrochemical Society*, vol. 142, no. 2, pp. 634–641, 1995.

[MAR 13] MARTIN M., AVERTIN S., CHEVOLLEAU T., *et al.*, "Fabrication of high-density Si and Si(x)Ge(1-x) nanowire arrays based on the single step plasma etching process", *Journal of Vacuum Science & Technology B: Microelectronics and Nanometer Structures*, vol. 31, no. 4, p. 041806, 2013.

[MOG 74] MOGAB C.J., LEAMY H.J., "Conversion of Si to epitaxial SiC by reaction with C_2H_2", *Journal of Applied Physics*, vol. 45, no. 3, pp. 1075–1084, 1974.

[MOR 09] MORKOÇ H., *Handbook of Nitride Semiconductors and Devices, Materials Properties, Physics and Growth*, Handbook of Nitride Semiconductors and Devices, Wiley, 2009.

[NIS 83] NISHINO S., POWELL J.A., WILL H.A., "Production of large-area single-crystal wafers of cubic SiC for semiconductor devices", *Applied Physics Letters*, vol. 42, no. 5, pp. 460–462, 1983.

[OLL 13] OLLIVIER M., Élaboration de nanostructures à une dimension à base de carbure de silicium, PhD Thesis, University of Grenoble, October 2013.

[OKA 84] OKADA Y., TOKUMARU Y., "Precise determination of lattice parameter and thermal expansion coefficient of silicon between 300 and 1500 K", *Journal of Applied Physics*, vol. 56, no. 2, pp. 314–320, 1984.

[OLL 13] OLLIVIER M., LATU-ROMAIN L., MARTIN M., *et al.*, "Si-SiC core-shell nanowires", *Journal of Crystal Growth*, vol. 363, pp. 158–163, 2013.

[SLA 75] SLACK G.A., BARTRAM S.F., "Thermal expansion of some diamondlike crystals", *Journal of Applied Physics*, vol. 46, no. 1, pp. 89–98, 1975.

[SMI 47] SMIGELSKAS A.D., KIRKENDALL E.O., "Zinc diffusion in alpha brass", *Transactions of the American Institute of Mining, Metallurgical and Petroleum Engineers*, vol. 171, pp. 130–142, 1947.

[STE 92] STECKL A.J., LI J.P., "Epitaxial growth of β–SiC on Si by RTCVD with C_3H_8 and SiH_4", *IEEE Transactions on Electron Devices*, vol. 39, no.1, pp. 64–74, 1992.

[STR 92] STRITE H., MORKOÇ H., "GAN, AIN, and InN–a review", *Journal of Vacuum Science & Technology B*, vol. 10, no. 4, pp. 1237–1266, 1992.

Chapter 4

[ALP 12] ALPER J.P., VINCENT M., CARRARO C., *et al.*, "Silicon carbide coated silicon nanowires as robust electrode material for aqueous micro-supercapacitor", *Applied Physics Letters*, vol. 100, no. 16, p. 163901, 2012.

[CHA 01] CHABERT P., "Deep etching of silicon carbide for micromachining applications: etch rates and etch mechanisms", *Journal of Vacuum Science & Technology B*, vol. 19, no. 4, pp. 1339–1345, 2001.

[CHA 08] CHAN C.K., PENG H., LIU G., *et al.*, "High-performance lithium battery anodes using silicon nanowires", *Nature Nanotechnology*, vol. 3, no. 1, pp. 31–35, 2008.

[CHO 12a] CHOI J.H., LATU-ROMAIN L., BANO E., *et al.*, "Fabrication of SiC nanopillars by inductively coupled SF_6/O_2 plasma etching", *Journal of Physics D: Applied Physics*, vol. 45, no. 23, p. 235204, 2012.

[CHO 12b] CHOI J.H., LATU-ROMAIN L., BANO E., *et al.*, "Comparative study on dry etching of alpha- and beta-SiCnano-pillars", *Materials Letters*, vol. 87, pp. 9–12, 2012.

[FRA 14] FRADETAL L., STAMBOULI V., BANO E., *et al.*, "Bio-functionalization of silicon carbide nanostructures for SiC nanowire-based sensors realization", *Journal of Nanoscience and Nanotechnology*, vol. 14, pp. 3391–3397, 2014.

[HOO 13] HOON CHOI J.I., Silicon carbide nanowires: from fabrication to related devices, PhD Thesis, University of Grenoble, 2013.

[JIA 11] JIA J., JU S.P., SHI D., *et al.*, "Electromechanical response of a SiC nanotube under local torsional deformation", *Journal of Physical Chemistry C*, vol. 115, no. 49, pp. 24347–24352, 2011.

[LAT 14] LATU-ROMAIN L., OLLIVIER M., "Silicon carbide-based one-dimensional nanostructures growth: towards electronics and biology perspectives", *Journal of Physics D: Applied Physics*, vol. 47, p. 203001, 2014.

[LEE 99] LEERUNGNAWARAT P., HAYS D.C., CHO H., *et al.*, "Via-hole etching for SiC", *Journal of Vacuum Science & Technology B*, vol. 17, no. 5, pp. 2050–2054, 1999.

[MAT 04] MATSUNAMI H., "Technological breakthroughs in growth control of silicon carbide for high power electronic devices", *Japanese Journal of Applied Physics*, vol. 43, no. 10, pp. 6835–6847, 2004.

[OLL 14] OLLIVIER M., LATU-ROMAIN L., SALEM B., *et al.*, "Integration of SiC-1D nanostructures into a nano-Field Effect Transistor", *Mater. Sci. Semicond. Process*, 2014.

[PLA 03] PLANK N.O.V., BLAUW M.A., VAN DER DRIFT E.W.J.M., *et al.*, "The etching of silicon carbide in inductively coupled SF6/O-2 plasma", *Journal of Physics D: Applied Physics*, vol. 36, no. 5, pp. 482–487, 2003.

[ROS 11] ROSAZ G., SALEM B., PAUC N., *et al.*, "High-performance silicon nanowire field-effect transistor with silicided contacts", *Semiconductor Science and Technology*, vol. 26, no. 8, p. 085020, 2011.

[ROS 01] ROSCHKE M., SCHWIERZ F., "Electron mobility models for 4H–, 6H–, and 3C–SiC", *IEEE Transactions on Electron Devices*, vol. 48, no. 7, pp. 1442–1447, 2001.

[SAD 11] SADDOW S.E., *Silicon Carbide Biotechnology: A Biocompatible Semiconductor for Advanced Biomedical Devices and Applications*, Elsevier Science, 2011.

[VIV 06] VIVEKCHAND S.R.C., RAMAMURTY U., RAO C.N.R., "Mechanical properties of inorganic nanowire reinforced polymer-matrix composites", *Nanotechnology*, vol. 17, pp. S344–S350, 2006.

[WAN 10] WANG Z.G., LI J.B., GAO F., *et al.*, "Tensile and compressive mechanical behavior of twinned silicon carbide nanowires", *Acta Materialia*, vol. 58, pp. 1863–1971, 2010.

[YAK 07] YAKIMOVA R., STEINHOFF G., PETORAL JR R.M., *et al.*, "Novel material concepts of transducers for chemical and biosensors", *Biosensors and Bioelectronics*, vol. 22, pp. 2780–2785, 2007.

[YAN 13] YANG Y., REN J.G., WANG X., *et al.*, "Graphene encapsulated and sic reinforced silicon nanowires as an anode material for lithium ion batteries", *Nanoscale*, vol. 5, pp. 8689–8694, 2013.

[YIH 97] YIH P.H., SAXENA V., STECKL A.J., "A review of SiC reactive ion etching in fluorinated plasmas", *Physica. Status. Solidi. (B)*, vol. 202, pp. 605–642, 1997.

[ZHA 07] ZHANG Y., HAN X., ZHENG K., *et al.*, "Direct observation of super-plasticity of β–SiC nanowires at low temperature", *Advanced Functional Materials*, vol. 17, pp. 3435–3440, 2007.

[ZHA 08] ZHANG T.Y., LUO M., CHAN W.K., "Size-dependent surface stress, surface stiffness, and Young's modulus of hexagonal prism [111] β–SiC nanowires", *Journal of Applied Physics*, vol. 103, no. 10, p. 104308, 2008.

[ZHO 06] ZHOU W.M., FANG F., HOU Z.Y., *et al.*, "Field-effect transistor based on β–SiC", *IEEE Electron Device Letters*, vol. 27, no. 6, pp. 463–465, 2006.

[ZHO 10] ZHOU R.L., WANG L., PAN B.C., "Elastic and melting properties of crystalline SiC nanotubes", *Journal of Physical Chemistry C*, vol. 114, no. 18, pp. 8199–8205, 2010.

Index

Other titles from

ISTE

in

Nanoscience and Nanotechnology

2014

BALESTRA Francis
Beyond CMOS Nanodevices 1
Beyond CMOS Nanodevices 2

SENEZ Vincent, THOMY Vincent, DUFOUR Renaud
Nanotechnologies for Synthetic Super Non-wetting Surfaces

TRIPON-CANSELIET Charlotte, CHAZELAS Jean
Nanoscale Microwave Engineering: Optical Control of Nanodevices

2013

NICU Liviu, LEÏCHLÉ Thierry
Micro-and Nanoelectromechanical Biosensors

2012

BELLEVILLE Marc, CONDEMINE Cyril
Energy Autonomous Micro and Nano Systems

CLAVERIE Alain
Transmission Electron Microscopy in Micro-nanoelectronics

LACAZE PIERRE Camille
Nanotechnologies: Concepts, Processing and Applications

2010

BALESTRA Francis
Nanoscale CMOS / Innovative Materials, Modeling and Characterization

QUERLIOZ Damien, DOLLFUS Philippe
The Wigner Monte Carlo Method for Nanoelectronic Devices: Particle Description of Quantum Transport and Decoherence

TAKADOUM Jamal
Nanomaterials and Surface Engineering

VODA Alina
Micro, Nanosystems and Systems on Chips: Modeling, Control and Estimation

2008

DAVIM J. Paulo, JACKSON J.
Nano and Micromachining

NOUAILLHAT Alain
An Introduction to Nanoscience and Nanotechnology

OUISSE Thierry
Electron Transport in Nanostructures and Mesoscopic Devices

2006

RIGNEAULT Hervé *et al.*
Nanophotonics

Printed and bound by CPI Group (UK) Ltd, Croydon, CR0 4YY

13/09/2026

14961017-0007